T0304445

Sourdough BY Science

Sourdough BY Science

Understanding Bread Making for Successful Baking

KARYN LYNN NEWMAN, PhD

The Countryman Press

An Imprint of W. W. Norton & Company
Independent Publishers Since 1923

All photography by Karyn Lynn Newman unless otherwise noted below:

Page 2–3: © BDMcIntosh/iStockPhoto.com; page 4–5: © Alina-kho/iStockPhoto.com; page 11: © DILSAD SENOL/iStockPhoto.com; page 13: © Barcin/iStockPhoto.com; page 16 (bottom left): © Floortje/iStockPhoto.com; page 22 (bottom left): © leonori/iStockPhoto.com; page 38: © Daria Dombrovskaya/iStockPhoto.com; page 85: © Veronika Roosimaa/iStockPhoto.com; page 112 (bottom left): © RyanJLane/iStockPhoto.com

Printed in China

Manufacturing through Imago
Book design by Nick Caruso
Production manager: Devon Zahn

The Countryman Press
www.countrymanpress.com

A division of W. W. Norton & Company, Inc.
500 Fifth Avenue, New York, NY 10110
www.wwnorton.com

978-1-68268-700-0

10 9 8 7 6 5 4 3 2 1

Dedicated to Jack, Orelia, and Jenya,
my three inspiring bread aficionados.

Contents

CHAPTER 9

PART FOUR

CHAPTER 10

Part One
Introduction

Part Two
How to Make
Artisan Sourdough
Part Three
The Artisan Bread
Recipes
Part Four
Traditional Breads
Using Your
Sourdough Starter
Part Five
Appendix

The smell of freshly baked bread is deeply intoxicating. Thumping a crusty, fragrant sourdough onto the table is a serious thrill—you know that it's going to be a great meal. But if you don't have easy access to a *really* good bakery, then that delicious fresh bread, still warm from the oven, is a rare treat. Learning to make excellent artisan bread for yourself is very simple in many ways, once you get the hang of it, and is actually a lot of fun (some might even say addictive). I discovered this by learning to bake artisan sourdough in a home kitchen myself, and I wrote this book to share what I learned so you can be up and running as a home artisan sourdough baker, *pronto*. You can have great bread, made the way you like it, whenever you want. Let's hear it for bread independence!

In my sourdough-baking journey, I have become familiar with the obstacles to making great bread, and I will help you avoid them. I have reviewed the bewildering array of possible gear, and I can help you choose the essentials for jumping right into successful sourdough bread making. I have studied bread making and tried the various and sometimes conflicting techniques and tips from a multitude of cookbooks, bakeries, flour producers, food scientists, and internet recipes and blogs. All along the way I pared down the bread-making process, keeping only the essential steps to creating fantastic bread at home with the least amount of effort.

It is regrettable that people sometimes meekly abandon sourdough baking after it appears to be an unsustainably time-consuming and effort-laden process. We are not going to let this happen to you! After all, you are probably living a very rich life already. Maybe you are busy with an intense career, a full family life, creative pursuits, socializing, exploring the outdoors, and so on. I want to help you fit artisan sourdough bread baking comfortably into your life and get you baking great bread right away. To do that, we will focus on a few key steps, techniques, and essentials, using mostly ingredients and equipment you already have on hand.

Demanding regimens for keeping the starter alive, long fermentations with hands-on requirements, and a perceived need for a lot of gear that must be selected, purchased, and stored, can frustrate timely success. Sure, you can make your own starter from scratch, knead your dough forever, do complicated steps involving lots of measuring, mixing, and timing, and buy fancy bakers or bannetons or proofing boxes. If you find that enjoyable, then by all means have fun with it. But you can make really great bread without all the fuss, and I've learned that if you want to be able to fit sourdough baking into your life in a sustainable way, then the best way to do that is to rely on a method that's fast, flexible, and foolproof. It can then be a delight to branch out from there if you choose.

Although you might want to make a great loaf of bread without taking time away from your other pursuits, a curious and intelligent person may still want to know why each step in the recipe is important. What would happen if you did it differently? You might want to

know how you can adapt a recipe successfully to your own preferred schedule, techniques, equipment, or ingredients. I will arm you with the information you need to understand how these steps and techniques affect your bread. Later you will be able to experiment to tweak your creations. And I give you reference points and context so you can understand the fermentation process and engineer it to your preferences. Using layperson's terms, I open the black box of dough development and show you what's inside. There's a reason why many cookbooks avoid doing this—it's a highly complex process and there are many variables that control how it goes. This complexity is why many great bakers will tell you to get your hands in the dough and learn to bake by experience. You most certainly don't need to understand fermentation to bake great bread, but you might want to because, you know, knowledge is power! Great bakers

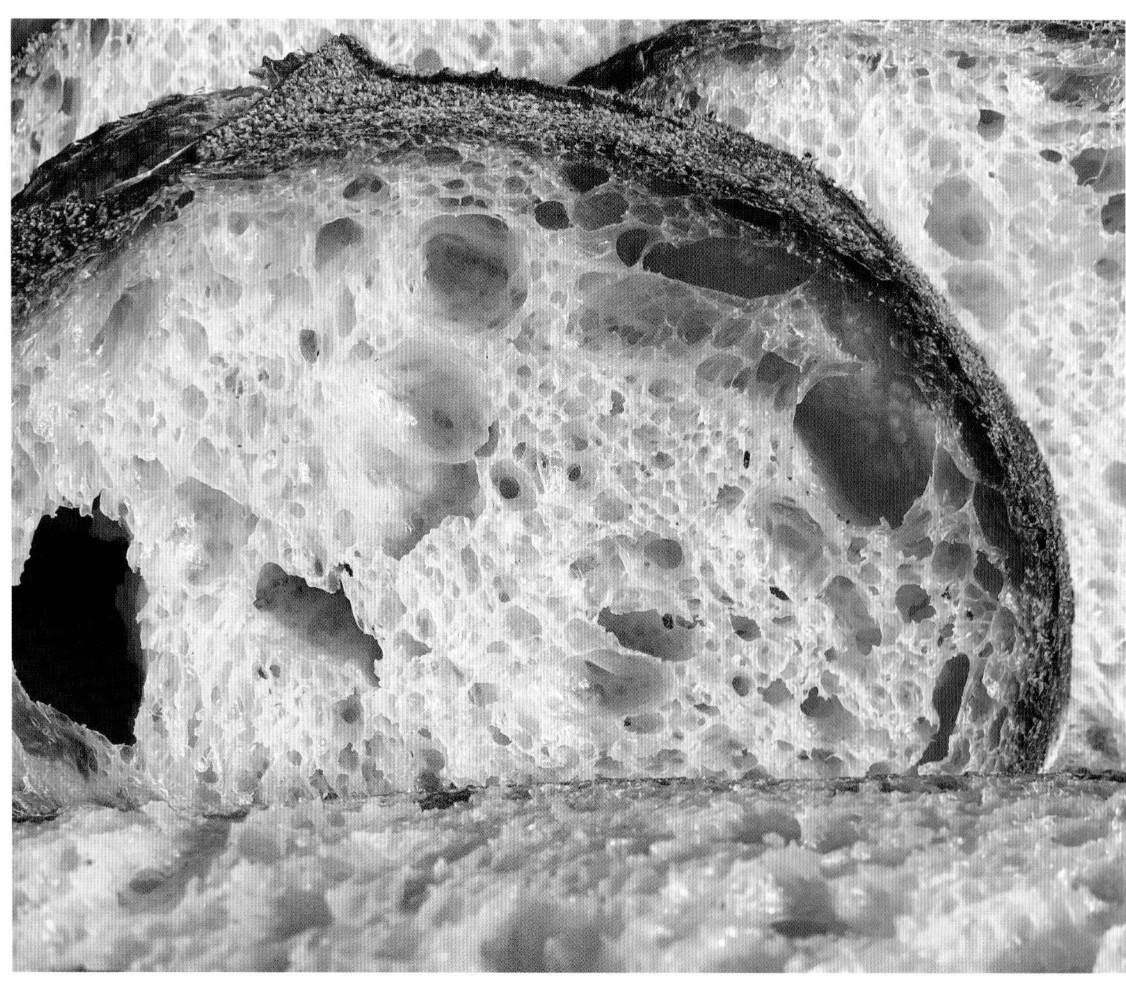

who don't know the scientific details and terminology have solid knowledge that is based on training and/or experience, which is a very powerful foundation. If you want to bake great bread without years of hands-on training and experience, you can use scientific knowledge of the bread-making process to get you started in the right direction.

Humans have used sourdough for thousands of years all around the globe. The diversity of methods employed and the breads that can be produced are infinite. Exploring these can be a joy, but it can also be confusing and overwhelming. What I present to you here is a manageable portfolio of ways to make artisan sourdough, and along the way I'll teach you what you'll need to know to make any recipe that you can dream up work for you.

CHAPTER 1

Getting Started

How to Use This Book

Read through Chapters 1 and 2, which will orient you and help you determine the items you'll need to have on hand for bread-making success. Next, read through Part 2 (Chapters 4 and 5), which explains everything you need to know to take care of your starter and bake an excellent artisan sourdough. If this seems like too much information right now, just focus on the instruction boxes and photos for each step. Then, start baking the Essential Recipes found in Chapter 6, referring back to the how-to and in-depth information as needed. Once you feel confident with the Essential Recipes, you can customize them using the ideas found in Chapter 7, or you can bake any recipe in the later chapters found in Part 3. In Part 5 of the book, you'll find a troubleshooting guide and a glossary, as well as a worksheet. You can make copies of the worksheet and use one each time you bake a recipe to help you keep track of what you did and how it turned out for future reference.

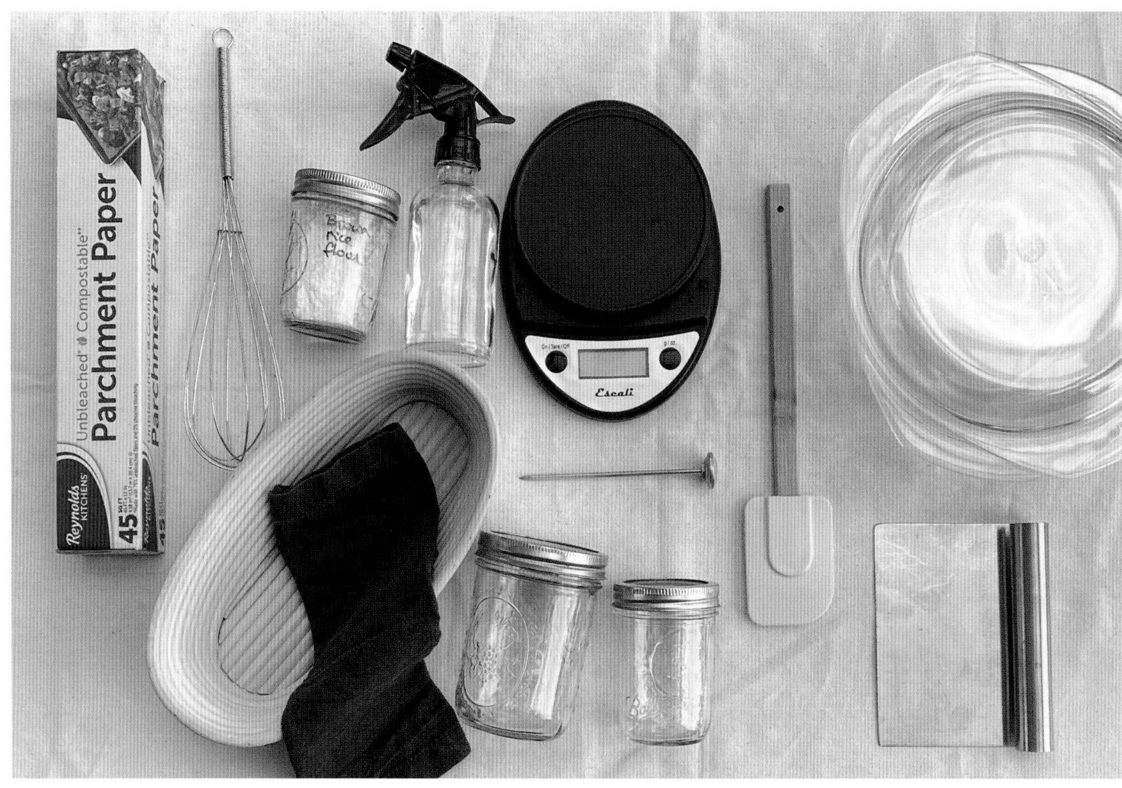

A few of my favorite sourdough baking essentials

Jump-Start Success

The main obstacles to quick success in beginning sourdough baking can be completely avoided by starting off with the essential equipment. You may already have most of them, or can get them for free or a pretty low cost. You will need:

- A fast-rising, **established starter**. You can ask your local bakery if they would share some starter (most do). Or you can order one online from, for instance, the 1847 Oregon Trail Sourdough Starter, which you can get for free from the website Celebrating Carl Griffith's 1847 Oregon Trail Sourdough Starter (carlsfriends.net). Or you can purchase a starter from King Arthur Baking Company (shop kingarthurbaking.com), Cultures for Health (shop.culturesforhealth.com), or Sourdoughs International (sourdo.com). Send away for one now, as it can take a couple of weeks to arrive. Beginning with an established starter gives you one less thing to suspect if you need to troubleshoot a recipe that didn't turn out as expected. It is fun to create your own starter from scratch or to try the one your neighbor just made. But if the starter doesn't double in volume within 3 hours at 74–80°F (23–27°C) after a refreshment, then it will not be able to raise the dough in a recipe and will lead to a lot of frustration and dense, gummy breads. I recommend getting an established 10-year-old or older starter even if you plan to make your own so you can see what a good starter is supposed to be like.

- A **draft-free location** that is a constant 75–80°F (23–27°C) for resting your dough. This is the optimum temperature for the yeast that makes the dough rise, and yeast responds poorly to temperature changes while it's raising dough. Some ideas:

    - A box on top of the refrigerator or near a computer tower (or in a cellar if your climate is hot).
    - An unlit oven with the light on.
    - An oven with a pilot light and the door held ajar.
    - A cooler with a bowl of hot water in it (or cold water if your climate is hot).
    - A box with a heating pad or hot water bottle (or bowl of ice if your climate is hot).

- An **instant-read thermometer** to check the temperature of your water, dough, and bread. This will allow you to estimate dough rising times in your kitchen, to optimize fermentation to your preferred schedule, to tell when your loaf is done baking inside, and to do other helpful things. I use a simple, inexpensive Taylor model 5989N.
- A **baking scale** with metric units (grams). I have a small, inexpensive Escali Primo P115C. Not only does a baking scale allow you to carry out recipes consistently and accurately, it also allows you to bake without any measuring cups, spoons, or sifters, which greatly minimizes cleanup. It also makes it very easy to halve or double recipes without laborious calculations.
- A roll of baker's **parchment paper** to use in these recipes. Parchment paper makes moving dough around easy and less stressful. Also, cleanup time is dramatically reduced (can you tell I'm not super into dishes?). It's compostable. It is not the same thing as waxed paper!
- **Rice flour** for sprinkling on your proofing basket, the container that holds your loaf while it's rising. This simple item can save you from spending 20 minutes sweating while trying to coax your fluffy little loaf out of its proofing basket without deflating it. For some reason, rice flour is way better than regular flour for this task, perhaps because it is slow to wet. You can find it in specialty or Asian food stores.

Also Recommended

- A **covered baking pan**, which will trap the steam coming from the dough as it bakes. This is a surefire way to get early success in baking a nice artisan boule.
    - This could be a cast-iron Dutch oven, a soup pot with nonplastic handles, a Pyrex- or ceramic-lidded casserole dish, or a graniteware roaster, to name a few. It must be lidded, oven-safe to 500°F (260°C), and hold 3 to 8 quarts (3 to 8 liters).
    - You could also use a baking sheet or baking stone with a metal mixing bowl or a large pot inverted over it, as long as there are no gaps.
    - A special Dutch oven or ceramic cloche designed just for bread are wonderful, but any of the covered bakers mentioned above will also bake great bread.
- An **old heavy pan**, preferably cast iron, to place on the bottom of your oven to boil water and to create steam throughout the oven while the bread bakes (needed only if you are not baking in a covered pan). Pans don't typically respond well to this treatment, getting scaly and rusty, hence the "old" qualifier.
- A **baking stone** is nice for pizzas and for creating a great bottom crust when used for breads.
- A **spray bottle** can be used for misting loaves with water just before baking.
- A **silicone spatula or dough scraper** is amazing for scraping the starter and dough out of containers and for mixing and folding sticky doughs so that the dough doesn't stick to your hands.
- A **bench knife**, preferably metal, but they also come in plastic. These are indispensable for dividing dough, tightening loaves, moving dough around without it sticking to your hands, and for rapid counter cleanup.
- **Mason jars and lids** for keeping your starter and for making levains. You can put the lid on without screwing it down tight for good sanitation while maintaining gas flow. You can see how high the starter has risen by placing a rubber band around the jar at the original height.
- A clear **glass casserole dish or bowl**, 3 to 4 quarts (3 to 4 liters), for bulk fermentations, or larger if you like to make two loaves at a time. It's great to be able to see through the container and look at the bubbles to understand how your fermentation is coming along. Clear plastic would also work, but plastic doesn't hold temperature as well as glass. A casserole dish comes with a handy lid for covering your dough, and it can also be used as your enclosed baking chamber when baking a boule.
- A **baking peel** is used to transfer loaves onto a hot baking stone. You could simply use a baking sheet with no rim on at least one side (my favorite), or a large cutting board in place of a peel.
- A **proofing basket**. This could be a banneton (a special willow basket made for bread) or another basket, colander, or bowl lined with a clean tea towel for proofing your loaves.

- **A way to keep dough from drying out**. Bowls of dough can simply be covered with a plate or lid. The parchment paper you will use later for baking can be laid over loaves, followed by an upturned mixing bowl, a slightly damp tea towel, or a puffed-up plastic bag. Plastic wrap is not recommended as it can stick to the dough.
- A working **oven** is required, but it does not have to be fancy. I bake in an old Wedgewood from the 1940s. Baking bread in an enclosed pot or casserole dish makes up for almost any deficiency your oven may possess. If your oven is really bad, with very uneven heat, your bread will benefit from being baked in a heavy baker, such as cast iron or ceramic. Something like a Lodge cast-iron combo cooker would be a great fix. Expensive enameled cast iron such as Le Creuset will bake very nicely, too, but will suffer aesthetically as the enamel may darken and brown.
- An **oven thermometer** can help you figure out if your oven runs too hot or cold (most do), and how long it really takes to preheat.
- A **standing mixer** for mixing and kneading dough is not needed for making breads from this cookbook. These are large and pricy. I have a KitchenAid, but I don't use it for sourdough recipes, except Wild Challah (page 236) and Rich Fluffy Dough (page 244). If you don't mind mixing dough by hand for 2 or 3 minutes, then you don't need one. You can still make great bread using these sourdough recipes because they are designed to capitalize on the natural tendency for gluten structure to develop spontaneously under the right conditions.

Measurements

Measuring flour by volume (cups) is tricky. You can lightly spoon sifted flour into a 1 cup measure and level it by scraping off the excess with a knife; it usually will weigh about 120 grams. But if you dip the cup into the flour and scoop it, it may weigh anything from 130 to 180 grams. This variability is highly problematic when following an artisan bread recipe, especially if it's your first time through it. Dough with a little extra flour or a little less flour is remarkably different to handle and requires different strategies to make good bread. A recipe with 4 cups of flour could wind up with as much as 720 grams, which is 50 percent more than the 480 grams called for, resulting in a lot of grief for the baker. You may be thinking that weighing flour is also problematic because flour in a drier climate weighs less than flour in a humid climate. This is true, but the differences are much, much smaller than those associated with measuring by volume. For example, the weight of flour in rainy Seattle versus sunny Phoenix is only 2 or 3 percent different than at milling.

For this and other reasons, the recipes in this cookbook are based on measuring flour by weight in grams. The standard generally used is 120 grams for 1 cup of flour, so that is the conversion standard used in this cookbook. **If you measure by volume, just make sure you lightly spoon sifted flour into the measuring cup and level it by scraping off the excess with a knife.**

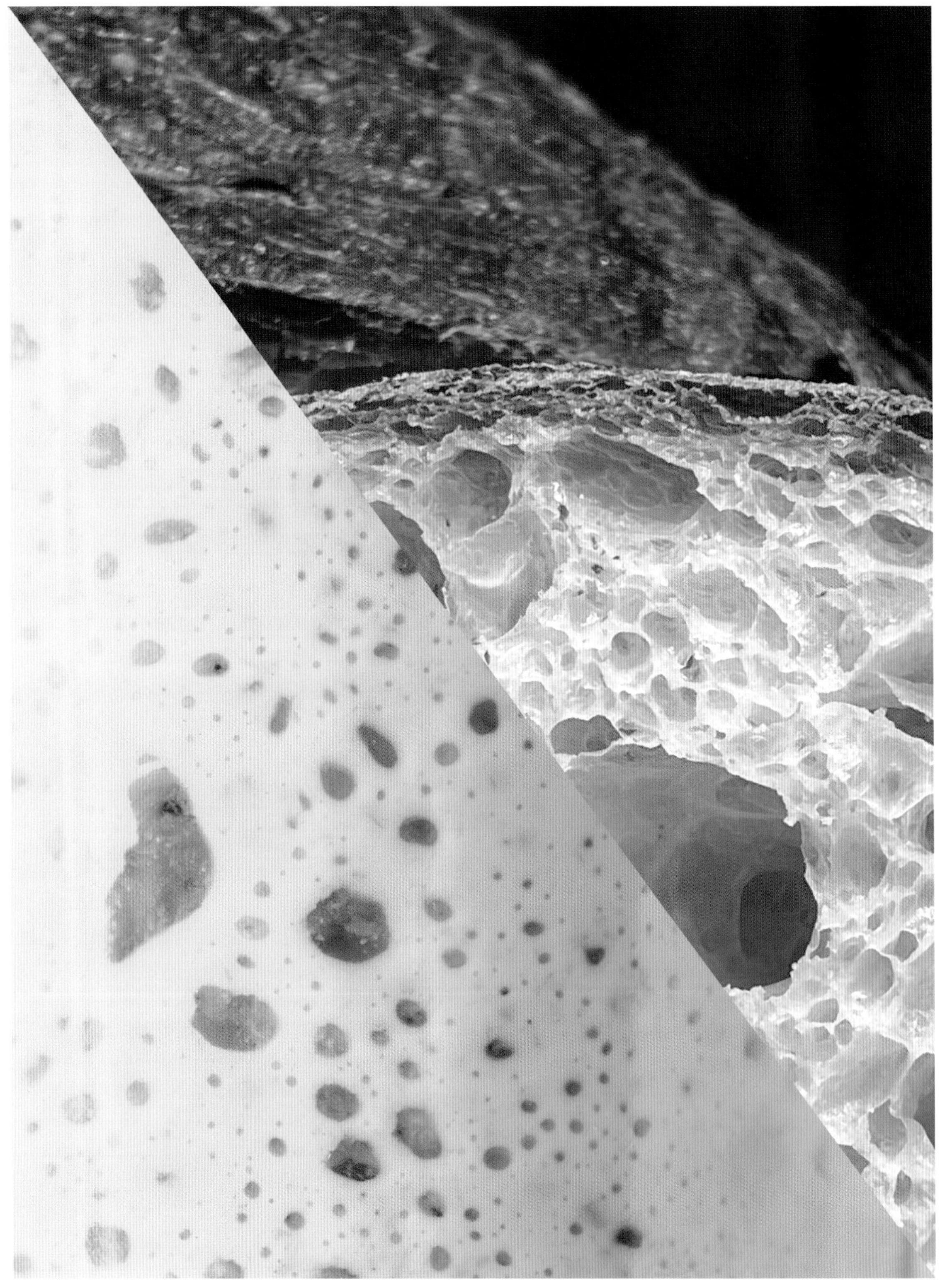

CHAPTER 2

Key Concepts

Sourdough Is Different from Regular Dough and Artisan Sourdough Is Even More Different

If you've made yeast bread before, you may find these recipes are unexpectedly unfamiliar. We are using wild yeast, not commercial yeast, so bread making is different. For one thing, wild yeast takes longer to raise dough than commercial yeast, which has been bred and optimized for the fastest rising. The longer rise is a good thing for artisan bread, really, because lots of flavor is produced by the sourdough yeast and bacteria during a longer rise, and the bread becomes more digestible and nutritious. Wild yeast performance can be variable. Every starter is different and the way the starter recently has been managed has a huge impact on the starter's performance in a recipe.

With certain modifications, it is entirely possible to use wild yeast in almost any bread recipe that calls for commercial yeast. There are many wonderful recipes and cookbooks available that cover sourdough recipes with low-hydration doughs, long kneading sessions, and punching down the dough. These recipes yield bread that is very similar to a commercially yeasted bread except that it has undergone bacterial fermentation. This cookbook is, for the most part, concerned solely with the creation of artisan sourdough breads, with their open crumbs and rustic crispy crusts and heavenly fragrance. When we do make an excursion into the world of breads that are typically made with commercial yeast and traditional kneading in Part 4, we use the techniques of artisan bread making in those recipes to the extent that they deliver a similar or better bread.

Artisan sourdough is a further departure from regular dough in that it is wet and sticky. These wet, or high-hydration, doughs yield a divinely open and chewy crumb, but they require a different handling strategy than regular dough. For example, we don't add flour to the dough to prevent it from sticking. Instead, we handle the dough using a brief touch with wet hands or implements. Gluten development does not come about by kneading, but by resting, along with some very short stretching and folding sets. Rather than punching down the dough at the end of its rising time, as is usually done, we preserve the bubbles in the dough while shaping the loaf. None of these techniques are particularly hard. In fact, I think they're much easier than those for regular bread. They just take a little practice before they become familiar.

What Is Sourdough Starter?

A sourdough starter is a community of wild yeast and bacteria living together in dynamic, symbiotic harmony and is used to raise dough and make bread. The yeast and bacteria need each other to survive, and their populations decline and increase independently in the starter according to how it is managed. Flour is their food. The yeast turn the flour into carbon dioxide gas bubbles and more yeast. The bacteria turn the flour into sugar for the yeast, tangy acids, and more bacteria. They both contribute lots of different flavor compounds to the bread. There are usually many species of bacteria and yeast in a sourdough starter.

How Exactly Does the Sourdough Starter Turn Flour and Water into Risen Dough?

The transformation of flour and water into a risen dough that can be baked into bread occurs during the bulk fermentation, which is a period of rest after the dough is mixed. A successful bulk fermentation is foundational to the finished bread's success. Two things must happen if you don't want your dough to bake into a brick:

1. Gluten must develop to make a strong dough that can hold bubbles throughout baking.
2. Yeast must produce carbon dioxide gas bubbles to make the dough puff up and rise.

Let's look at these two processes more closely.

Gluten Makes a Strong Dough

Gluten development occurs when the gluten proteins found in flour join to each other and become aligned in long, parallel strands and sheets. This gluten structure is what gives dough the ability to stretch (extensibility) and to spring back (elasticity). It also gives the dough the ability to hold bubbles and to rise stably. When dough is first mixed, the proteins join with each other randomly, in a big snarly mess like a bowl of spaghetti, and the dough feels very stiff. Once the gluten structure is developed and organized, the dough feels stretchy and elastic.

There are two ways the proteins can realign into a nice, organized gluten structure: spontaneous and mechanical. With enough time and enough water, the proteins arrange themselves spontaneously into organized gluten because it is the most energetically favored arrangement of wet gluten proteins. In other words, they want to be organized, and given the opportunity they will. Or the dough can be kneaded (the mechanical method), which helps the gluten to organize rapidly without requiring as much water in the dough. Think of it like salt dissolving in water. Over time, a spoonful of salt will spontaneously dissolve in a glass of water, but if you stir it around you can really speed up the process.

Since wild yeast raises bread more slowly than commercial yeast, there is already going to be a period of waiting no matter what. There's no clear advantage to kneading, and long kneading does actually have some downsides, such as causing bleaching of the carotenoids by oxidation. For this reason, many sourdough recipes call for extra water in the dough (high-hydration dough) and minimal or no kneading, often replacing it with a very quick and easy technique called stretch and fold.

The key point is that you can get the necessary gluten structure several different ways.

- A very slow fermentation with no dough manipulation at all (requires a lot of water in the dough).
- A medium-slow fermentation with some stretch and folds (requires a moderate amount of water).
- A medium-fast fermentation with a little kneading and stretch and folds (requires a normal amount of water).
- A very fast fermentation with a lot of kneading (this requires commercial yeast and a normal amount of water).

Carbon Dioxide Gas Puffs Up the Dough

Carbon dioxide gas is produced mainly by the yeast that is growing and multiplying in the dough. This gas is responsible for puffing up the dough with bubbles. When the yeast eats the food available in the flour, it makes gas. As time goes on, the number of yeast cells doubles, then doubles again and again until the dough is full of yeast, which are making gas, causing the dough to be puffy with bubbles. The yeast cells can go on doubling and making gas until they run out of food in the dough. Bread is best when the yeast has puffed up the dough but hasn't run out of food before the loaf goes into the oven.

In general, yeast grows fastest at about 74–78°F (23–26°C). As you lower the dough temperature, it grows more and more slowly. Also, yeast cells grow by doubling in number. Starting with a high number of yeast cells will fill the dough with yeast faster than starting with a low number of yeast cells. Bacteria, which are also in the starter, grow very quickly at 90°F (32°C) and above, leading to rapid fermentation and flavor development, but they don't do much to puff up the dough.

The key point is that you can get carbon dioxide gas to puff up the dough several different ways.

- A very slow fermentation using a small amount of starter and/or a low temperature of 40–60°F (4–16°C).

- A medium-slow fermentation with a medium amount of starter and/or a medium temperature of 65–70°F (18–21°C).
- A medium-fast fermentation with a large amount of starter and/or an ideal yeast temperature of 74–78°F (23–26°C).
- A very fast fermentation with commercial yeast and an ideal yeast temperature of 74–78°F (23–26°C).

Finally, because gluten development and carbon dioxide production are both happening in the dough, the bread maker needs to match the conditions of the bulk fermentation to accommodate both of these processes so they are complete at the same time. Slow fermentation no-knead recipes, such as those in Chapter 9, call for small amounts of starter and low to medium temperatures so there's time for the spontaneous gluten structure to develop. Fast recipes, such as those in Chapter 8, call for large amounts of starter (in the form of a levain, which will be explained in Chapter 5), ideal yeast temperatures, and more stretch and folds to develop the gluten before the food runs out in the dough.

Stretch and Fold

There is no need to knead a well-hydrated sourdough if you just stretch the dough and fold it, then let it relax a few times. This is a simple and ridiculously fast and easy way to encourage gluten development in a well-hydrated dough. Each set takes less than 1 minute of hands-on dough manipulation followed by a rest period of 15 to 60 minutes. Typically, a recipe will involve two or three sets of stretch and folds separated by rest periods. When the dough is stretched, the gluten proteins are encouraged to align and join with neighboring proteins in an organized pattern while letting go of connections that are unfavorable. During the rest period, the proteins let go of more connections that don't fit in the organized pattern and find new ones that do. Then in the next stretch they can find new connections and let go of unstable ones. Eventually, lots of the proteins are bound together in an organized network that gives the dough strength and stretch.

Handling Sticky, High-Hydration Dough

High-hydration sourdough is very sticky (especially right after mixing and during the bulk fermentation), which could cause a lot of frustration and difficult cleanup. Luckily, we have ways around this. Once you get the hang of it there is very little sticking or mess. The best strategy to use depends on where you are in the process. In general, though, it pays to handle the dough as briefly as you can, like it's a hot potato. It only takes a couple of seconds for it to glue itself to whatever it touches that is not wet, floured, or greased. Don't be tempted to flour everything liberally, however, because your sourdough bread will not turn out the way it's supposed to if extra flour is incorporated into it, and it will still stick to everything and make a big mess anyway.

In the early stages of the recipe, flour isn't very effective to prevent sticking, but water is. Wet your fingers and implements

frequently as you handle your dough. I keep a mister or a bowl of water nearby. Lightly mist the countertop before turning out your dough, and use damp hands and a damp bench knife to divide and preshape the loaf.

When it's time to shape your loaf, a little flour is usually the most effective way to prevent sticking, but the less you use, the better. Raw flour folded inside the loaf results in a tough layer in the finished bread.

You will proof the loaf top side down in its proofing basket. You may choose to use flour (especially rice flour), seeds, or bran to keep the dough from sticking to the proofing basket before you turn it out to bake it. Or you can proof the dough top side up with parchment paper underneath (being sure to keep the dough covered so it doesn't dry out). The parchment paper can be used to lift the dough from the proofing basket to move it directly to the baking surface and can remain under the dough as it bakes.

Knowledge Is Power

Now that you understand the parameters that lead to successful bulk fermentation and dough handling, you can assess how a recipe will work for you. You can also figure how to make the bulk fermentation optimal for the conditions of your kitchen, your schedule, and your preferences.

- Prefer not to knead at all? You'll need to make sure your fermentations happen at lower temperatures or without large amounts of starter (see Chapter 9).
- Maybe you live where it's 90°F (32°C) without air-conditioning. You'll want to incorporate some extra stretch and folds or kneading toward the early part of your bulk fermentations to make sure the gluten is developed before the food runs out in the dough.
- Maybe you have a tight schedule and need the dough to be ready fast. If so, you may prefer a warmer, fast fermentation and so you will want to stretch and fold or lightly knead the dough early on.
- Maybe you have a cold kitchen. You'll need a longer amount of time for bulk fermentations, so fewer stretch and folds will be needed.

Amber
Durum
Whole
wheat
Buckwheat
Rye
Spelt
UNBLEACHED

CHAPTER 3

Flours Used in This Cookbook

Flour choice is a very personal thing. Bakers have their own parameters based on what is available and affordable, their flavor and nutrition goals, cultural beliefs, traditions, and habits. All these factors influence which kind of flour they bake with. The recipes in this book will work with all-purpose wheat flour found at any store, from the enriched, bleached, white flour widely available throughout the United States to unbleached organic flour from boutique mills. Using flour from a company known for its dedication to a high-quality, fresh product will enable more consistent results and better flavor. The recipes in this book were test baked with a variety of flours with good results. I can personally recommend flours from King Arthur Baking Company, Bob's Red Mill, Central Milling Company, Community Grains, and Giusto's, although there are many, many other sources of high-quality flours.

The freshness of flour greatly affects the flavor and nutrition of breads. White flour is designed to be very shelf stable, but it can absorb odors from its surroundings over long periods of time and take on a stale smell. Whole grain flour can become rancid, tasting bitter or having "off" flavors or odors. It can be advantageous to purchase whole grain flour in small amounts from a fresh source, such as a local mill or a store with rapid turnover, or to buy flour packaged with an expiration date stamp. Whole grain flour should be kept in an airtight container and refrigerated if not used within a few weeks. For the absolutely freshest and best-tasting flour, you can grind your own from whole grains using a home flour mill, although it is an investment. Whole grain berries can be purchased for low cost and stay fresh for eons in cool, dry conditions. For those who choose this route, it becomes as simple (and as addictive) as grinding your own coffee beans.

Flours from Common Wheat, *Triticum aestivum*

ALL-PURPOSE FLOUR: Standard white flour. It is milled from soft and/or hard wheat and has been sifted to remove the bran and germ. Levels of gluten protein (which gives the dough strength) vary among brands and are not always reported on the package. It comes bleached or unbleached and can contain various additives. It yields breads with a delicate flavor and a soft, light crumb.

BOLTED FLOUR/SIFTED FLOUR/EXTRACTED FLOUR: Whole grain flour sifted (a.k.a., bolted or extracted) to remove the larger pieces of the bran, leaving behind the endosperm, the wheat germ, and some smaller pieces of bran. These flours are used to make essentially whole wheat breads that are able to rise higher due to a lack of large bran pieces. Bolted flours can vary tremendously from one another depending on the size of the sifting screen and the coarseness of the grind. The extraction rate tells you what proportion of the flour went through the screen (for example, an 85 percent extraction flour has had 15 percent of its weight removed through sifting). A "high-extraction" flour is one that has retained much of its weight and is therefore closer to a whole wheat flour than a white flour (which is generally about 72 percent extracted).

BREAD FLOUR: White flour milled from wheat that has a higher amount of gluten proteins, which yields a stronger dough for bread making. Results in a tall loaf and chewy crumb. Bread flour absorbs more water than all-purpose flour and can support better loft in breads with whole grains, weak gluten grains, or heavy ingredients such as nuts. All-purpose flour can be substituted. To replicate bread flour, use 3 grams (1 teaspoon) of vital wheat gluten (see the following Vital Wheat Gluten entry) in place of 3 grams (1 teaspoon) of all-purpose flour per 120 grams (1 cup) of the recipe.

TIPO 00: A very soft and fine white flour from Italy that is used in making tender pizza crusts and breads. All-purpose flour can be substituted.

VITAL WHEAT GLUTEN: Gluten protein processed from wheat flour, which can be used as an additive when extra gluten strength is desired in a recipe. Useful when substituting all-purpose flour in a recipe calling for bread flour, or to increase the strength of doughs made from weak gluten flours or that have a lot of heavy ingredients such as nuts. Overuse results in chewy bread. Available in most large grocery stores.

WHITE FLOUR: A general term that can be used to refer to any refined wheat flour missing the bran and germ, or it can be used to refer to all-purpose flour specifically.

WHITE WHOLE WHEAT FLOUR: Similar to regular whole wheat flour, but milled from wheat that lacks red pigment in its bran. White whole wheat flour has a delicate and sweet flavor without the tannic bitterness of red whole wheat, and it makes breads that are lighter in color.

WHOLE WHEAT FLOUR: Whole grain flour milled from soft or hard wheat and sold with some or all of the bran and the germ. Whole wheat flour has more flavor, color, and nutrition than white flour. It must be handled differently than white flour in dough because of the interaction of the bran particles with the gluten network and because it ferments faster than white flour. It is commonly made from red wheat but whole grain white wheat flour, which has a lighter color and a more delicate flavor, can be found at specialty retailers. Whole grain ancient wheat flours can be substituted but may affect dough handling.

Flours from Ancient Wheat Species

DURUM: *Triticum durum*, an amber-yellow-colored wheat grain that is very high in protein. It makes a rather delicate gluten structure that doesn't like to extend, and so it requires gentle, minimal handling. Durum is commonly found either as a whole grain flour, as semolina (a coarse ground flour lacking bran and germ that is used to make pasta), or as semola (a very finely ground flour lacking bran and germ that is sometimes labeled rimacinata). This grain makes a lovely soft and satiny lemon-yellow dough and has a distinctly delicious flavor. Durum flour absorbs more water than common wheat. Freekeh is durum that has been harvested while still green by burning the chaff from the wheat. It is pale green in color and has a wonderful smoky taste, but it is best used only as an accent flour.

EINKORN/FARRO PICCOLO: *Triticum monococcum*, an ancient variety of wheat from the Fertile Crescent that spread to Central Europe. It has a slightly golden color and a sweet, nutty taste. It forms a weak gluten structure and does best in recipes with minimal dough handling.

EMMER/FARRO: *Triticum dicoccum*, an ancient wheat from Europe and Asia also referred to as farro (sometimes einkorn and spelt are also called farro). Emmer, spelt, and einkorn are expensive to produce because their hull is difficult to remove as compared to the other free-threshing wheats. Evidence of its domestication and use by humans dates back to 7000 BC. Frequently eaten as a cooked whole grain, its red-brown flour can be substituted for common wheat, but it has low-quality gluten.

KHORASAN: *Triticum turanicum*, an ancient wheat from the Middle East and Central Asia with a buttery, nutty taste. It has giant-sized grains and is sold under the trade name Kamut. It makes a delicate gluten structure and smooth dough.

SPELT/DINKEL/FARRO GRANDE: *Triticum spelta*, an ancient wheat from Central Europe, also called dinkel or farro grande. It has a nutty but mild, sweet taste and makes a more tender crumb than common wheat. It has a unique gluten structure that is highly extensible but with very little strength, and therefore it cannot produce a very tall loaf without support. Spelt doughs benefit from light, gentle handling. Spelt flour absorbs less water than common wheat.

Non-Wheat Flours

AMARANTH: Technically a seed and not a grain, amaranth was used by the Aztec and Inca societies. Its malty, nutty, slightly sweet flavor is a fun addition to a loaf of bread and brings with it the essential amino acid lysine, which is low in many grains.

BARLEY: Barley has a wonderful nutty flavor, but it has very weak gluten that cannot raise bread well. Its flour can be substituted for up to 25 percent of the wheat flour in a bread recipe. Malted, or sprouted, barley has a distinctive sweet aroma. As long as it hasn't been overheated, malted barley contains a group of active enzymes that jump-start yeast fermentations by making maltose available early in the fermentation process and adds a wonderful malt aroma to the bread. Barley is usually chosen when making diastatic malt powder. Pearled barley has had most of the bran removed and likely is missing most of the underlying aleurone layer, where the enzymes reside.

BUCKWHEAT: Buckwheat, a pseudocereal, lends a savory flavor and pretty, dark flecks to waffles. It pairs well with other strong flavors.

CORN: Corn flour lends a sweet, creamy flavor to breads. When coarsely ground, it can be used to dust a baking peel or pan to keep dough from sticking.

DIASTATIC MALT FLOUR: Generally ground from sprouted barley or wheat, this flour provides a hefty dose of the amylase enzymes responsible for liberating maltose from the starches found in flour. Maltose, a deliciously fragrant sugar, is favored by yeast and when plentiful it can boost yeast growth and activity. Amylase enzymes are naturally present in wheat flours, but amounts vary between batches of wheat. Many mills add malt to their standard wheat flours to achieve consistency, though not all do. This flour adds a subtle malt aroma and extra height to bread, but it can be left out of a recipe with no dramatic changes to the outcome.

MILLET: This grain from Africa has a mild, sweet flavor, and millet flour adds a delicate crumble and extra nutrition. Used whole, it adds a nice crunch to breads.

OATS: Oats can be added to breads as flour, rolled oats, or as soaked whole oat groats. They impart a creamy flavor, velvety texture, and heartiness to the bread.

RICE: Rice flour has excellent antisticking capabilities when used to line a proofing vessel.

RYE: Rye hails from Central Europe and has a sweet grassy or floral scent that is very distinctive. Rye flour is marketed as either light rye, which has had the germ and bran removed; dark rye, which is whole grain or nearly whole grain and boasts a violet-blue cast; or pumpernickel, which is whole grain and may also be coarsely ground, depending on the brand. Rye is high in amylase enzymes, which enable rapid fermentation, and pentosans, which make rye dough feel sticky. Possessing a delicate gluten due to a shortage of the gliadin protein, a rye dough is best handled gently and minimally. Structure in an all-rye dough comes more from the starches and pentosans than the gluten.

Sprouted Grains

Any raw true grain can be sprouted, then dried and ground into flour for nutrition, enzymatic, or flavor purposes. Note that oat groats must be specially labeled for sprouting as most of what are sold are not raw.

How to Make Sprouted Grains

Choose grains that are from a fresh source, and are whole and intact. They should not be pearled, "quick," or rolled, but the straw-like husk, or hull, does need to have been removed.

Choose a container that leaves plenty of room for the grains to spread out and swell. I like to use a 2-quart jar (2 liter) for 1 cup of grain (190 grams). You'll also need a piece of cheesecloth and a band for the jar top.

Place the grains in the jar and add water to rinse them, then drain the water through the cheesecloth lid. Cover the grains with fresh water and rest at 65–75°F (18–23°C) for 12 hours.

Drain the water, rinse the grains gently in the jar, and drain the jar again completely so no grains are sitting in water where they might begin to ferment or mold. To ensure no water remains, you can set the jar upside down or at an angle over a bowl. Alternatively, if your kitchen is humid, you can transfer the grains to a canvas bag, which you rinse with water and hang for ideal drainage and airflow.

Repeat the rinse-and-drain process every 8 to 12 hours until you see little white roots growing, about one to five days.

Rinse one last time and drain very well.

At this stage, the sprouted grains can be eaten fresh in salads, added into bread dough, stored in the refrigerator for a day or two, or dried for grinding or long-term storage.

How to Dry Sprouted Grains

In order to preserve the enzymatic activity and the best flavor of sprouted grains, they must be kept cooler than 100°F (38°C). To dry them, use a food dehydrator, an oven barely warmed then turned off, or a warm, dry spot in your home with good air circulation to prevent mold. If enzymatic activity is not important and you just want to make flour from sprouted grains, any temperature below 200°F (93°C) will be fine.

Spread grains in a single layer on a screen and allow them to dry completely. When dry, they should be hard and weigh the same as the grains you started with. Drying can take somewhere between a few hours and a day, depending on your setup and your weather.

Dried grains can be ground in a grain mill. A food processor, coffee grinder, blender, or spice mill will also grind them, but you may need to pause frequently to avoid overheating the appliance or the flour. The finer the grind, the more enzymatic activity the flour will impart to the dough.

Part One
Introduction

Part Two
How to Make Artisan Sourdough

Part Three
The Artisan Bread Recipes

Part Four
Traditional Breads Using Your Sourdough Starter

Part Five
Appendix

CHAPTER 4

The Care and Keeping of a Sourdough Starter

We begin at the beginning—a sourdough bread can only ever be as good as the starter used, so we want to make sure you know how to keep a top-notch starter with the least fuss and effort. The following information assumes that you have already gotten your hands on an established starter from a trusted source. Are you interested in making your own brand-new starter? Check out the instructions at the end of this chapter, Create Your Own Starter from Scratch (page 48).

A culture that has been frequently refreshed with new flour and water over the past few days has a high proportion of actively dividing yeast. That culture will bubble, rise, and work well in a recipe. A culture that is or has recently starved has a higher proportion of bacteria, because the yeast cells have begun dying off and going into a sort of hibernation state. In this condition, a starter will rise slowly, may produce a liquid called hooch, and will work unpredictably in a recipe. But it can be brought back to recipe-ready condition by the simple act of refreshing it one or more times. By changing the feeding and temperature regimen, you can move a starter from bubbly to hibernating and back as often as you like.

A note on measuring: You will get consistent results by using a scale to measure your starter. A starter's volume is capricious, changing all the time, so if you opt to measure by volume be sure to stir it down vigorously before measuring. When choosing a container for your starter, remember that it can rise to fill two to four times its original volume, so leave a lot of headspace and keep the lid loose so gas can escape.

There are many, many effective routines for keeping a healthy sourdough starter. The key is to choose a routine that fits comfortably into your life and provides a starter that makes bread you like. The information in this chapter will help you establish an ideal routine for your baking adventures. For purposes of illustration, I will first describe how I manage my starter and how the routine supports my baking objectives.

My starter, Viola; just after being refreshed

Viola; ready to store in the refrigerator

My Routine

I keep 60 grams (¼ cup) of my favorite starter, Viola, in a half-pint jam jar in the refrigerator. I like to refresh the starter right after I use some to bake. One to three times a week, I remove 30 grams and use it in a recipe. After I mix the recipe, I remove 10 grams out of the starter jar and discard it so that 20 grams of starter are left. Next, to bring it back up to 60 grams. I mix 20 grams of tap water and 20 grams of flour in the starter jar. I leave the starter jar on the counter next to my recipe in progress. If I'm planning to bake the next day, I keep it there overnight to let the starter rise. If not, a few hours later, when I can see many bubbles in the starter and the starter is beginning to rise, I store it in the refrigerator where it slowly continues rising.

If I haven't baked in a week, a note on my calendar reminds me to refresh the starter in a 1:2:2 proportion (proportions are explained in a following section). Using this routine, my starter is always recipe-ready right out of the jar from the refrigerator, meaning it is risen and a spoonful will float in water. This gives me flexibility for spontaneous baking projects.

Periodically, when refreshing the starter I transfer it to a fresh clean jar, largely for aesthetic reasons.

Sometimes life intervenes and poor Viola isn't refreshed for several weeks. Or sometimes I am concerned about the health of the culture (maybe there's hooch, or it's inexplicably runny, and so forth). In these cases, I discard nearly all the starter in the jar and refresh it 1:7:7. It usually takes 12 to 24 hours before the starter rises, but then it's ready to bake with successfully.

If the starter has been grown in the refrigerator for several refreshments in a row, I grow it on the counter after the next refreshment to help keep the bacterial populations from getting too low.

I also have a stash of dried starter (see page 46) in the freezer in case any disaster should occur.

This routine suits me because I can whip up a loaf or two of bread on a whim anytime the mood strikes, and I can consolidate refreshing the starter with baking (that is, when I've got all the supplies out already). I always have recipe-ready starter available, yet I don't generate a lot of discard or wasted flour. The weekly refreshment reminder, along with the frozen dried starter, function as insurance against disasters so I can afford to have a relaxed attitude toward the whole endeavor and still make consistently fantastic bread.

As you can see, there can be a lot of flexibility in caring for your starter. I hope that my example can help you come up with a routine that works for your life and baking ambitions.

Keeping a 60-gram (¼-cup) Starter in a Small Jar

For any recipe in this book, having 60 grams (¼ cup) of starter on hand is plenty. For some reason, people commonly keep much larger starters, but I find the amount of discard generated troubling. I'd rather be baking bread with all that flour! I also prefer the streamlined half-pint jar that fits so nicely in my refrigerator to the giant crock. There is no reason to keep a starter much larger than what you would use in a recipe. If you do need more, you can easily make it and use it a few hours later. Keeping a smaller starter ensures that your culture is regularly grown anew, and that in turn will help prevent issues with over-soured or over-fermented breads.

What to Do with the Discard

Whenever you refresh the starter, there is some leftover old starter, or discard. You can throw it away or you can use it in a sourdough discard recipe from this book or one of the many found on the internet. In my starter management routine, I generate only small amounts of discard, which I save in a pint jar in the refrigerator. When I have a cup (230 grams) of it, I make Wild Waffles (page 262).

Proportions and Hydration Levels

Starters are kept at a particular hydration level. In this cookbook, we use a 100 percent hydration level: the weight of the water is 100 percent of the weight of the flour in the starter. That means for 1 part of water, we add 1 part of flour by weight. A 1 part to 1 part proportion is expressed as 1:1. When refreshing the starter, let's say we want to take 1 part starter and add 1 part new water and 1 part new flour. This is a 1:1:1 proportion of starter:water:flour. Sometimes we want to add more fresh water and flour to our starter. Then we might use a 1:2:2 proportion of starter:water:flour or even a 1:7:7 proportion.

The water temperature should be room temperature, around 60–80°F (16–27°C). The starter will rise faster at the warmer end of that range and slower at the cooler end.

In general, if you are refreshing your starter frequently, such as every day or two, a 1:1:1 proportion is fine. One-third of the weight of the new mix is old starter (which has living yeast and bacterial cells but also their wastes) and two-thirds is fresh food. If you refresh less frequently, such as every week, there will be more wastes in the old starter, so you will want to add less of it. A 1:2:2 proportion will help keep the starter healthy in that case. If your starter has been neglected for more than a week or two, a 1:7:7 refreshment will help it snap back to its former vigor. It will take a longer time than usual to grow back, but it will be nice and vigorous once it does. Don't ever worry that you are using too little of the old starter—there are billions of yeast and bacterial cells in one tiny gram of sourdough starter.

Following are instructions for the most frequently used refreshments of a 60-gram (¼-cup) starter (devoid of math in case I just gave you spiral eyeballs).

Refresh Your Starter 1:1:1

This is a good proportion to use if you are refreshing your starter every day or so.

1. Measure into a clean half-pint jar or keep behind in the original jar (discarding any excess starter):

20 grams starter (4 teaspoons)

2. Add and stir until smooth:

20 grams water (4 teaspoons)

3. Add and stir until smooth:

20 grams flour (3 tablespoons)

4. Mark its height and let it rest at room temperature.

5. When it is risen by half you can store it in the refrigerator or keep it at room temperature.

6. When it is at or around its peak height (usually two to three times its original height) you can bake with it.

7. When it has begun to deflate you should refresh it again soon.

Refresh Your Starter 1:2:2

This is a good proportion to use if you are refreshing your starter only once or twice a week or if it makes hooch or deflates too soon when you use a 1:1:1 regimen.

1. Measure into a clean half-pint jar or keep behind in the original jar (discarding any excess starter):

    12 grams starter (2½ teaspoons)

2. Add and stir until smooth:

    24 grams water (1 tablespoon plus 2 teaspoons)

3. Add and stir until smooth:

    24 grams flour (3 tablespoons plus 1 teaspoon)

4. Mark its height and let it rest at room temperature.

5. When it is risen by half you can store it in the refrigerator or keep it at room temperature.

6. When it is at or around its peak height (usually two to three times its original height) you can bake with it.

7. When it has begun to deflate you should refresh it again soon.

Refresh Your Starter 1:7:7

This is a good proportion to use if you are refreshing your starter after it has been neglected for more than a week or any time you are worried about its health.

1. Measure into a clean half-pint jar or keep behind in the original jar (discarding any excess starter):

    4 grams starter (1 teaspoon)

2. Add and stir until smooth:

    28 grams water (2 tablespoons)

3. Add and stir until smooth:

    28 grams flour (¼ cup minus 1 teaspoon)

4. Mark its height and let it rest at room temperature. It will take much longer than usual to rise.

5. When it is risen by half you can store it in the refrigerator or keep it at room temperature.

6. When it is at or around its peak height (usually two to three times its original height) you can bake with it.

7. When it has begun to deflate you should refresh it again soon.

How to Test If a Starter Is Ready to Use in a Recipe after Being Refreshed

If you drop a spoonful of ready starter in water, it will float. If it doesn't float, it is either too soon to use it or it has deflated and needs refreshment again. It is better to use this float test than to count on a particular amount of time, as so many variables change the amount of time it takes a starter to rise and then deflate. These variables include temperature, proportion of starter used when refreshing, and the particular characteristics of the individual starter.

Where to Store Your Starter and How Often to Refresh It

Do you bake almost every day or enjoy the routine of daily starter refreshment?

- Store it on your countertop and refresh 1:1:1 every day when you normally bake. Or . . .
- Store it in the refrigerator, taking some when you need it and refreshing each time you bake or when it's running out. At low temperatures, the starter uses up the food in the flour much more slowly, but the starter needs refreshment at least once a week to be ready for use in baking. After refreshing, leave it out until it has begun to rise, then refrigerate.

You can use the starter directly in a recipe as long as it floats. If it seems to run out of food too soon (makes hooch), change to a 1:2:2 or 1:3:3 refreshment routine, or put it in a slightly cooler place.

Do you bake once or twice a week or month?

- Store your starter in the refrigerator and refresh it 1:2:2 when you bake. Put a reminder on your calendar to refresh it once a week in case you are too busy to bake.
- If it seems to run out of food in less than a week (makes hooch), change the refreshment proportion to 1:3:3.

You can use the starter directly in a recipe as long as it floats. Storing your starter in the refrigerator without periodically allowing it to rest at warmer temperatures will favor yeast in the starter and breads will be less sour. For more sour bread, refresh at a warm room temperature of 80°F (27°C) a few times in a row every so often. Following the instructions for Revive a Neglected Starter (see page 45) allows you to do this without generating discard.

Not baking for a while?

- You can dry your starter and store it in the freezer long term (see page 46).
- You can do a heavy refreshment, such as 1:7:7, and leave it in the refrigerator for several weeks or maybe longer (every starter is different).

In both of these cases, you will need to refresh the starter, possibly more than once, before you can bake with it.

How to Tell If You Can Bake with a New or Revived Starter

Refresh the starter 1:1:1, mark its height, and let it rest at 74–80°F (23–27°C). If it doubles in height within 3 hours, it will perform well in a recipe.

If it doesn't, refresh again (and maybe again) until it does. If it never does, you probably need a different starter if you want to bake successfully.

Why Not Bake with a Slow-to-Rise Starter?

Whatever is happening in your starter jar after refreshment is similar to what will happen in your bowl of dough. It could take a long and unpredictable amount of time to raise the dough if you use a slow-to-rise starter. This presents an opportunity for the bacteria in the starter to eat all the food before the yeast can raise the dough very far, so your dough could become over-fermented without ever having risen very far. If it's only a little bit slower than normal to rise and the starter has performed well in the recent past for you, it's probably okay to use.

Revive a Neglected Starter

If you are using an established starter, it is very, very hard to actually kill it by neglecting it in the refrigerator. Like a phoenix rising from the ashes, the starter will come back to life, but it takes time. Hooch (clear or colored liquid) is perfectly fine and normal in a hungry starter, but check for black or fuzzy mold or brightly colored or slimy bacterial patches. If you find any, throw away the starter and seek a new one.

1. Stir vigorously until it's a uniform consistency.
2. Measure into a clean half-pint jar:

   5 grams neglected starter (1 teaspoon) or less

3. Add and stir until smooth:

   10 grams water (2 teaspoons)

4. Add and stir until smooth:

   10 grams flour (4 teaspoons)

5. Mark its height and let it rest at room temperature until it has risen to double its height. It should be within 24 hours, but proceed to the next step after 24 hours anyway.
6. Add and stir until smooth:

   25 grams water (2 tablespoons)

7. Add and stir until smooth:

   25 grams flour (¼ cup)

8. Mark its height and rest at 74–80°F (23–27°C).
9. Note how long it takes to double in height. It should double by 24 hours.

10. If it can double in 2 or 3 hours, you can use it to bake. (If it hasn't made any bubbles at all after 24 hours, it may be dead.)

11. If it's rising slowly, keep 20 grams (4 teaspoons) and perform the regular 1:1:1 refreshment-and-rest routine one or more times, allowing it to fully rise each time, until it can double in 2 or 3 hours.

What Is Hooch Anyway?

Hooch is a liquid full of acids and alcohols made by the yeast and bacteria in your starter as they transition their metabolism from eating the abundant starches and sugars found in fresh flour to alternative food sources. It is full of flavor—not that I'm suggesting you drink it—and it contains such things as acetic and lactic acids, alcohol, and other flavorful compounds. It can be mixed back into your starter or thrown away. It's kind of like that packet that comes in a ramen package. Some dump it in their soup, some dump it in the trash. It's all a matter of taste.

Dry Your Starter to Freeze It or Mail It

1. Refresh the starter 1:1:1 or 1:2:2 and mark its height.

2. At peak (usually two to three times the height), spread thinly, about ⅛ inch (3 millimeters) thick, on parchment paper

3. Dry at 85–90°F (29–32°C)—no higher!—for 3 to 6 hours until bone dry. Use a food dehydrator or your barely warmed oven turned off. Use a thermometer to ensure you don't kill the starter by going over 95°F (35°C).

4. Crush into a fine powder. Use a mortar and pestle, spice grinder, rolling pin, or a blender.

5. Place in an envelope or bag. Store in an airtight jar to freeze for long-term storage. It must stay totally dry to survive.

6. After storing, test that you can actually revive it (see Revive a Dried Starter below). Every wild culture is different.

7. Put in small plastic bags and mail to friends and family.

Revive a Dried Starter

1. In a half-pint jar stir together:

> **1 gram dried starter powder (½ teaspoon)**
>
> **10 grams lukewarm water (2 teaspoons)**

2. Let stand 5 minutes.

3. Add and stir until smooth:

> **9 grams flour (1 tablespoon)**

4. Rest at room temperature until it gets fairly bubbly (usually 12 to 18 hours).

5. Add into the jar and stir until smooth:

> **20 grams water (4 teaspoons)**

6. Add into the jar and stir until smooth:

> **20 grams flour (3 tablespoons)**

7. Mark its height and rest until it peaks (usually 8 to 24 hours).

8. Keep 20 grams (4 teaspoons) and perform a 1:1:1 refreshment and rest at 74–80°F (23–27°C).

9. It should now be doubling in height within 2 to 3 hours at 74–80°F (23–27°C), but you may have to refresh it one or two more times before it can double fast enough.

10. Once it has doubled in 2 to 3 hours at 74–80°F (23–27°C), you can bake with it.

How to Get a Starter to Make Bread More Sour

If your bread is not turning out as sour as you like, you may need to give your starter a couple of refreshments at warmer temperatures to allow the bacteria, which make the acids, to increase their population. Yeast will grow best at 74–78°F (23–26°C) but will still grow pretty well at lower temperatures. Bacteria really need warmth to grow significantly (this is one reason why refrigeration preserves food). To increase the population of bacteria in your starter, give it some time to grow at 85–90°F (29–32°C) after it has risen a little. Be careful not to overdo it. Once the bacteria get too numerous, they may inhibit the rising action of the yeast possibly by eating all the food, by creating too much acid, or by eating the gluten network. A starter that made hooch, smells nose-scorchingly acidic, or has transformed into a runny liquid has gone too far at the warm temperature. You will have to experiment because every starter is different. You can always fix things by refreshing, so experiment all you like. Giving the loaf a cold retard (see page 161) after shaping is a good way to increase the sourness because some bacteria produce acetic acid under these conditions. Some starters lack bacterial species that can churn out acetic acid. If these strategies coupled with a cold retard don't make your bread more acidic, think about getting a different starter.

How to Get a Starter to Make Bread Less Sour

If your bread is turning out too sour, you may need to give your starter a couple of prompt refreshments at cooler temperatures to allow the yeast to increase their population. Yeast grows best at 74–78°F (23–26°C), but it will still grow pretty well at lower temperatures, just more slowly. Bacteria really need warmth to grow significantly (this is one reason why refrigeration preserves food). To increase the population of yeast in your starter, keep it at 40–70°F (4–21°C) after it has risen a little. Refresh it as soon as it has begun to deflate. After two or three refreshments, your starter should have a higher population of yeast and a lower population of bacteria. It should start to smell more yeasty, less acidic. You will have to experiment because every starter is different. You can always fix things by refreshing, so experiment all you like. Be sure that when baking, you are not resting the dough at temperatures above 80°F (27°C), as this would allow the bacterial population to increase.

Create Your Own Starter from Scratch

Although I strongly recommend embarking on sourdough bread baking with an established starter, once you know the ropes, you will no doubt be curious to try making your own starter. It's fun!

When grains are harvested from the fields, they are covered by millions of wild yeast and bacteria cells from the environment. They include the kinds of bacteria and yeasts that like to eat flour, and these are the ones you want in your starter. When flour is milled, these microorganisms wind up in the flour. Microorganisms are more plentiful in whole grain flour because the bran, which is on the outside of the kernel, is included. When the flour is mixed with water, the microorganisms begin to grow. What follows is a succession of different species dominating the culture, culminating in a set of yeast species that are suitable for raising dough. A collection of lactic acid bacteria will come along, too, benefiting from and helping the yeasts. Once this population of yeast and bacterial species is established, it is very stable. You will have created your very own microbial ecosystem!

What to Expect

While you are looking at your jar of flour paste and hoping it turns into a starter, you may be curious about what's going on in there day by day. Here's how the succession of species in the jar generally unfolds.

During the first 1 or 2 days, the conditions in the culture are not actually great for wild yeast, which prefer an acidic environment, so they will just quietly hang out. Instead, bacterial species that like low-acid environments begin to grow and end up producing acid as a waste product.

After a couple of days you may see some bubbling as evidence that these pioneer bacterial species are growing. Eventually, they produce so much acid that they cannot grow anymore. But the acid-loving species needed for bread making are now provided with an environment in which they can grow.

You may see the starter do nothing for a day or two at this juncture. The species that like an acidic environment are slowly beginning to grow and further acidifying the ecosystem. This includes the kind of yeast that raise bread as well as lactic acid bacteria.

After being quiet for a few days, all of a sudden your starter may become active again with lots of bubbles and may rise dramatically. At this point you have successfully grown baking yeasts!

During the next few refreshments, the baking yeasts will become dominant in the starter ecosystem, along with their lactic acid bacteria friends. Usually this entire process takes between 5 and 14 days.

Create a New Starter

1. In a pint jar stir together:

> **30 grams whole grain flour—rye, wheat, or spelt (¼ cup)**
>
> **30 grams water (2 tablespoons)**

2. Cover loosely and rest at 65–70°F (18–21°C) for 48 hours, stirring occasionally to oxygenate the mixture, which will help the right kind of pioneer species to grow. There should be some bubbles by the end of 48 hours.

3. Add and stir until smooth:

> **30 grams all-purpose flour (¼ cup)**
>
> **30 grams water (2 tablespoons)**

4. Cover loosely and rest at 65–70°F (18–21°C) for 24 hours.

5. Transfer 20 grams (4 teaspoons) to a half-pint jar and refresh 1:1:1 once every 24 hours until it begins to rise at least double its height in the jar, usually after 3 to 10 more days. Discard the extra starter without using it in a recipe.

6. Test the starter to see if it is strong enough to bake with. Refresh the starter 1:1:1, mark its height, and rest at 74–80°F (23–27°C). If it doubles in height within 3 hours, it will perform well in a recipe and you can safely use its discard.

Give your starter a name and continue daily refreshments for as long as you can to help the starter become stable and to perform consistently before storing it. Plan to refresh it daily for at least a month, continuing longer if growth and/or baking performance is inconsistent. It can take several years until a starter has established a dominant strain of yeast, and each refreshment encourages this stabilization process.

When using your new starter, expect some inconsistencies and evolution in performance. To achieve the earliest consistent performance, establish a convenient routine of refreshment (timing/type of flour/temperature/hydration) and stick to it as best you can. This will help the starter settle into a stable community. If you want to experiment with changes to the routine, you can always make an offshoot starter (by transferring a few grams of the starter to a second jar and refreshing that jar independently) and also keep your original going with the original routine.

CHAPTER 5

Stages of the Recipe, Step by Step

The main stages of sourdough bread making are described in this chapter. We will go through them step by step, with in-depth explanations followed by a succinct basic set of instructions that you can quickly locate and refer to while baking.

1. Preparing the starter or levain
2. Measuring and mixing the dough
3. Bulk fermentation
4. Shaping the dough into a loaf
5. Proofing the loaf
6. Baking

Preparing the Starter or Levain (2 hours to overnight)

Refresh Starter or Make Levain

The recipes in this book begin in one of two ways. In the levain recipes, such as in Chapters 6 and 8, you will first build a levain for the recipe. A levain is essentially a special large batch of starter that will all end up in the dough and be baked into the bread. In the slow fermentation recipes, such as in Chapter 9, you will build all the dough at the beginning of the recipe. In both cases, you will be using some of your starter. Before beginning these recipes, your starter should have been refreshed within the previous 3 to 24 hours if kept at room temperature, and within the last week if kept in the refrigerator. It should be bubbly and light, not totally deflated, gummy, runny, or producing liquid (called hooch), which indicates the food in the flour is gone. At its last refreshment, it should have doubled in size within the usual amount of time for your starter at the temperature you stored it. If not, it may not be able to raise your bread dough very well. If you refresh the starter, 1:1:1, using 90°F (32°C) water and keep it between 74–80°F (23–27°C), it should be ready to use in a few hours. To refresh your starter, see page 42.

A ready levain has risen at least twice its height in the jar . . .

. . . and has a domed top full of bubbles.

You may be tempted to use your unrefreshed starter to "save time," but your bulk fermentation may end up taking much longer since the yeast in an unrefreshed starter is kind of hibernating. Once unrefreshed starter is mixed into new flour and water, there is a lag phase before yeast is able to begin the rapid growth required to raise dough. If bulk fermentation takes extra time because of yeast lagging, gluten structure can break down due to bacterial enzyme action and turn your dough into a sour, soupy mess that won't rise well in the oven.

If the recipe begins with making a levain, you have more latitude with using a starter that hasn't been refreshed as recently because the levain itself offers the yeast an opportunity to pass through its lag phase.

The levain recipes here begin with growing a slow levain overnight in a 1:2:2 proportion of starter to water to flour. However, if you prefer a different schedule you could grow a fast levain and use it the same day. Simply change the proportions to 1:1:1 (for example, 50 grams each of starter, water, flour), warm the water to 90°F (32°C) before mixing, and let it rest at 74–80°F (23–27°C). It should be ready to use in 2 to 4 hours. A fast levain is best prepared with a recently refreshed starter.

To make the levain, measure out the starter into a clean jar or bowl. Add the water and whisk together with a fork until smooth. Add the flour and stir until completely mixed. Complete mixing is key because sourdough organisms cannot move around to find the flour. Cover the jar loosely and place a rubber band around the jar to mark the height after mixing so it will be easy to see how much the levain has risen later on. Rest at the temperature called for until it is risen at least twice its height but hasn't yet deflated more than half an inch (1 centimeter) or so. If no temperature is specified, then any ambient temperature 60–80°F (16–27°C) is fine, and the dough will rise faster at the warmer end of that range and more slowly at the cooler end.

Measuring and Mixing the Dough (10 minutes)

Measure Ingredients

For these recipes, your bowl needs to hold at least 3 quarts. A heavy material such as glass or ceramic will hold the temperature more steadily than metal or plastic, and a steady temperature helps the yeast perform its best. A glass bowl allows you to see the bubbles forming in the dough.

Use a scale! This simplifies everything and improves the probability of success enormously. It ensures you can reproduce a bread you had really liked and makes it very easy to double or halve a recipe. An added bonus: There will be no measuring cups to wash. Familiarize yourself with your scale's tare function to make things easier.

Note: If you measure flour by volume, you must fluff up the flour before measuring. This can be done by sifting or just inverting

Getting ready to mix up some dough during morning coffee

Floating levain or starter signifies a good population of yeast is ready to make your dough rise.

the container a couple of times, then spooning the flour into the measuring cup and sweeping off the top. If you do this, flour weighs 120 grams per cup, which is the standard used to convert grams to cups in this book. If you scoop flour with the measuring cup, you will probably put too much flour in your dough and it will be too dry. A little extra flour can make a very big difference.

Unless you are using a slow fermentation recipe, warm up your bowl and the water to get things off to a good start; 90°F (32°C) is as warm as you should go or you may kill the starter. Once mixed, the dough will end up with a temperature something like 74–80°F (23–27°C). If your dough is much colder than the recipe calls for after mixing, it will take longer for the next steps to happen. For slow fermentation recipes, the water should be no warmer than 60–70°F (16–21°C).

To measure the ingredients, place your mixing bowl on the scale and tare it. Pour in the flour until you get the correct weight. Tare the scale. Continue with the rest of the dry ingredients, remembering to tare the scale after each one. Whisk the flours together and set aside. Place a bowl or measuring cup on the scale, tare it, and pour in the water at the temperature called for until you get the correct weight. Continue with the other wet ingredients and the levain or starter, taring the scale after each addition. Whisk the wet ingredients together completely before mixing with the flour.

If you do measure by volume, you may need to adjust the flour or water in the recipe. Wait until 30 minutes after you mix the dough to decide whether to adjust, because it takes time for the flour to absorb the water.

Mix Dough

I highly recommend mixing the dough with a silicone spatula or dough scraper. It takes about 3 minutes and is not very hard to do for most recipes in this cookbook. To make bulk fermentation more efficient, whisk the water and starter together thoroughly with a fork, then add this to the flours. This ensures that the starter is well dispersed in the dough.

You can find many sourdough procedures that have you mix with your hand, but then they proceed to have you knead for quite a while to force the gluten to form

Beginning to mix

Midway (looks so dry!)

and to transform the dough into a less-sticky state. That's a perfectly fine way to do things, but I prefer to let gluten form spontaneously while the dough is sitting around fermenting anyway, so that's how the recipes in this book are written. A few recipes in this book with lower hydration doughs do call for kneading, but for most of the recipes, we will not be kneading the dough.

You can use your hands if you want, but these sourdoughs are sticky, wet creatures, and a third of the dough may end up irretrievably stuck to your fingers. Scrape it off as best you can. Do not use flour to "help" mix the dough without it sticking—by the time you have added enough flour so that the dough doesn't stick, you will have destroyed the painstakingly calculated hydration level required for spontaneous gluten formation and open crumb. Over-floured dough may bake into a tight-crumbed, dense brick.

If you prefer, dough can be mixed in your stand mixer with the dough hook. Add the flours to the bowl before the liquids and use a low speed ("level 2" on a KitchenAid). The

Ready for the autolyse

dough likely will not form a clean ball due to the high hydration of these doughs. Just mix the dough enough to combine completely. There is no need to knead in the mixer unless the recipe calls for it.

Once the dough is mixed, the sourdough organisms will be able to access all the flour and the bulk fermentation will begin, so you should make a note of the time and temperature of the dough.

To mix the dough, stir with a silicone spatula or dough scraper until all the flour is wet. This should take 2 to 3 minutes total. There is no benefit to further mixing once all the flour is evenly wet. Make a note of the time and temperature.

Keeping the dough within the temperature called for helps ensure a successful fermentation that is completed on schedule.

Autolyse (30 minutes)

Once all the flour is wet, over time the water will soak completely into the flour particles, including the gluten proteins. This process of soaking the flour in water to activate gluten formation is termed autolyse—it's a French word that refers to the autolysis (breakdown) of gluten that occurs when flour is fully hydrated—and it happens faster at higher temperatures. You can see and feel the dough go from a shaggy, tough ball to a softer, smoother, flatter, wetter ball at the end of a good autolyse. If the dough is still stiff after the autolyse time period, check the temperature to be sure it's in the range called for in the recipe. It will need more time if it is colder than the recipe calls for. Wait until the dough is ready before proceeding (and try to find a warmer situation or your bulk fermentation will also take longer than expected). See the Troubleshooting section (page 276) for help adjusting the temperature of the dough.

This step actually speeds up the recipe because gluten cannot form until the gluten proteins are fully hydrated (wet). The autolyse is designed to cause the most rapid hydration of the gluten proteins. In this cookbook, we soak the flour in water with the starter already added to get the bulk fermentation started, saving a little time. In recipes with whole grains, the autolyse has the added advantage of softening the bran particles. Once they are softened, they do less damage to the gluten network when the flour is handled, so the bread can rise higher and have a lighter crumb. Salt, large amounts of sugar, fats, or acids slow down hydration of the gluten proteins, so these are added afterward unless a very long bulk formation is used, such as in a slow fermentation recipe.

For the autolyse, cover the dough and let it rest at 74–80°F (23–27°C) until it has relaxed and softened and has become stretchable and less sticky than it was right after mixing.

Add Salt

Don't forget the salt! It is needed for flavor and gluten structure, and it also keeps the fermentation from happening overly fast. I like to put the salt jar on top of the plate covering my dough bowl during the autolyse so I am less likely to forget it.

Why not just add the salt along with everything else? Because salt interferes with the hydration of the gluten proteins. If you do choose to add it earlier, your bread will be fine, but it may take longer to get a nicely developed gluten network. In the slow fermentation recipes, we add the salt at the beginning because the dough has plenty of time to form a gluten network during the long bulk fermentation.

Spatula folds mix the dough and develop the gluten.

To add the salt, sprinkle the salt on the surface and poke and pinch it deep into the dough with your wet fingers. Then thoroughly mix the dough in the bowl with the silicone spatula by folding dough from the edges of the bowl toward the center, going around the bowl several times until it becomes harder to fold the dough, the salt is incorporated, and the dough is in a loose ball shape. This takes usually 15 to 30 folds. If the dough is of a lower hydration, you may find it easier to use your damp hands to finish mixing in the salt and to form a ball. Flip the ball in the bowl and cover.

Bulk Fermentation: When Flour Transforms to Risen Dough (2 to 4 hours)

Bulk Fermentation

This is the main event during which the transformation of flour and water into risen dough occurs! There's not much for you to do as the work is being done by the sourdough organisms and good old natural chemistry, but that doesn't mean you should totally check out.

Once the dough is bubbly and seems lighter and fluffier, it won't be long until it is able to jiggle in its bowl when you shake it. When it is jiggly like a bowl of jelly, the dough is ready to be shaped. If you miss the dough becoming jiggly and go past this phase, things can begin to go south rapidly, so pay attention toward the second half of the bulk fermentation and be prepared for the following steps in the recipe. I will arm you with a posse of tests you can use to figure out how close to ready your dough is during the bulk fermentation.

If life intervenes during your bulk fermentation and you cannot attend to your dough, just move it into the refrigerator. This will dramatically slow down the fermentation. When you're ready to give it your attention again, simply bring it out of the refrigerator.

Temperature is critically important in this step because the two most important things that need to happen will slow way, way down if it's not warm enough. Recall that these two things are:

1. The growth of yeast results in gas bubbles that puff up the dough.
2. The gluten network forms, which holds those bubbles in place so they can make it all the way through baking.

However, if it's too warm—much higher than 80°F (27°C) or so—the bacteria can grow very fast while the yeast grows pretty slowly, which is problematic. Excessive bacterial fermentation can overpower and derail the growth of yeast and the stability of the gluten network. In other words, try to find a way to rest the dough at the temperature indicated in the recipe for best success.

You will notice in the recipes that a wide range of time is given for how long the bulk fermentation should be expected to take. The shorter time corresponds to the warmer end of the dough temperature range called for and the longer time corresponds to the cooler end. In addition to temperature, many variable factors can slow down or speed up a fermentation: The individual starter's characteristics and its recent management are key factors, and there are also dozens of seemingly minor factors that would be hard to catalog, from flour choice to weather to dough handling. Even if you've made the recipe before the same way, it may take a different amount of time. The range is meant to provide a guideline for scheduling purposes, and under most circumstances the recipe as written will be ready in that time window. The baker needs to check on things periodically and to use the cues detailed in this book to determine when the dough is ready.

During bulk fermentation, check on the dough periodically. Perform stretch and fold sets if the recipe calls for it. Take notes on the dough's characteristics (smell, bubbles, amount risen, stretchiness, strength) and note the time and dough temperature to be sure it is within the range called for in the recipe. Not much will be happening at the beginning, but then things tend to happen faster and faster toward the end.

Stretch and Fold

During the bulk fermentation, the gluten proteins are slowly floating around in the water of the dough, bumping into and joining up with each other and forming a gluten network. At first, the proteins just link up with each other in a tangled mess, and so the dough is a tough, shaggy ball. Over time they stretch out into their naturally preferred organized network, and so the dough feels satiny and less sticky. In order for this to happen, they need to come near each other to link up and they need to let go of poorly bound links so that they can rebind in a better link.

Spontaneous gluten formation does occur, but it takes a long time, so some recipes speed things up with a quick 30-second-long set of procedures called stretch and fold. During stretch and folds, we are giving the proteins a little tug to help them let go of poorly bound partners and to link up with a new, better one. We gently stretch the dough so that the proteins get moved toward new potential binding partners. Of course, we don't want to lose the organized network that has been built thus far, so we never pull on the dough to the point where it is tearing.

Now that the gluten proteins have been tugged around, we need to give them time to relax so that they can shed the weak, disorganized bonds and link up with new, organized ones. It's a law of nature that molecules in water move faster at warmer temperatures, and gluten proteins are molecules. So check the temperature of the dough and if it's not at the temperature needed for the recipe, move it to a location with an ambient temperature closer to that needed for the recipe. Consult the Troubleshooting section (page 276) for ideas.

After a set of stretch and folds, there is a rest period during which the ball of dough relaxes and flattens before the next set. This rest period is usually 15 to 30 minutes, but it can be an hour if the dough is cold. If the dough hasn't relaxed from its ball shape and the dough is at the correct temperature, then it is either too soon for another stretch and fold or the gluten network is already strong enough and no further manipulation is required. Once the dough feels strong and elastic and loosely holds its shape for 15 to 30 minutes, you can let the dough continue to rest and rise undisturbed for the remainder of the bulk fermentation. During this period of undisturbed rest, the dough will rise and develop bubbles.

Loosening the dough from the bowl

Grabbing a hunk

Stretching it up

Folding it over toward the other side of the bowl

To perform a set of stretch and folds, get a bowl of water and a silicone spatula. Dip the spatula in water and loosen the dough in the top "north" quarter of the bowl. Dip all your fingers in the water to dampen them. Grab the hunk of dough you just loosened with both hands and pull it up until it resists, then fold it over toward the "south" bottom side of the bowl. Rotate the bowl and repeat this on the south side, then repeat on the "east" and "west" sides until all four sides have been stretched and folded. If the dough seems kind of rubbery or shows any sign of tearing, stop stretching. If the dough still seems soft and stretchy after all four sides are done, continue with another one or two times around until it feels strong and resists. Now flip over the ball of dough. This is considered one set of stretch and folds. Follow up with a period of rest (15 minutes or longer, until the dough relaxes again). Pay attention to specific instructions in recipes with unusual flours, as variations may be needed.

Note: some bakers prefer to use a variation called a coil fold. To coil fold, loosen all the dough in the bowl with the spatula. Lift all the dough using your wet fingers (palms up). The dough will stretch down. Lower it back into the bowl, folding it back on itself in a coil or spiral (from the side view). To complete a set of coil folds, turn the bowl a quarter turn and repeat, continuing until the dough resists stretching.

Bubble and Volume Test

Rising is a key feature of dough development. It results in bubbles forming in the dough, both large and small, depending on the type of dough. It also results in an increase in volume of the dough. Both of these can be seen most easily in a see-through, straight-sided container.

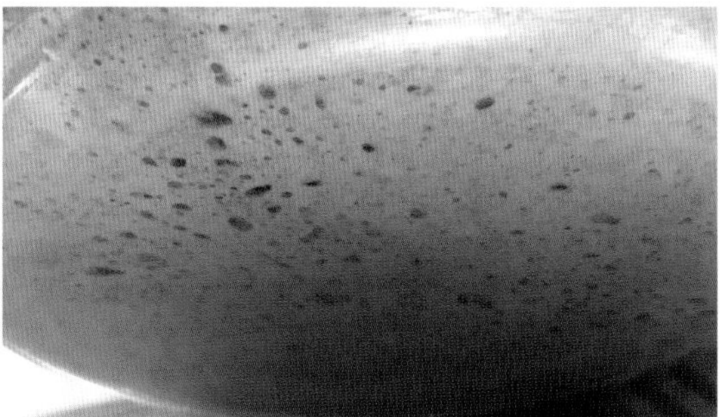

Ready dough has bubbles visible throughout.

To check bubbles and volume, note the height of the dough just after mixing and mark where it should be when the dough has risen the amount called for in the recipe. Check the height periodically during bulk fermentation and note bubbles forming in the dough. When it is getting close, it will rise faster and bubbles will grow quickly. Until the dough meets the expected height—as called for in the recipe and is full of bubbles—it must continue resting at the temperature called for in the recipe.

Ready dough has risen to at least double its original volume.

Windowpane Test

Eventually your dough will have a good enough gluten network; but how can you tell? The windowpane test is very useful for judging the quality of the gluten network. If your dough has whole grain flour, the size of the windowpane in the finished dough will be smaller. The more whole grain in the dough, the smaller the windowpane you should expect. Also, some types of grain have weaker gluten and will not be able to give a big windowpane, so check the recipe. For lower hydration dough, you may have to really stretch it, slowly, to get the windowpane.

To check the windowpane, grab a hunk of dough with wet fingertips and pull it up. If it makes a thin sheet several inches wide that you can see light through, then it passes the windowpane test. If it tears right away or is too thick to see through, it fails the test. If you are checking during the autolyse, failure means the autolyse should continue longer. If you are checking during the early part of bulk fermentation, failure indicates you should continue with folding sets.

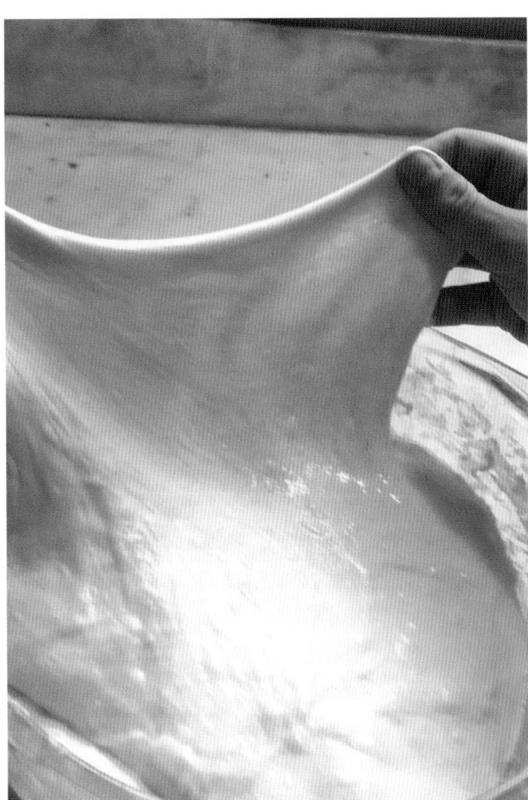
A windowpane in white flour dough

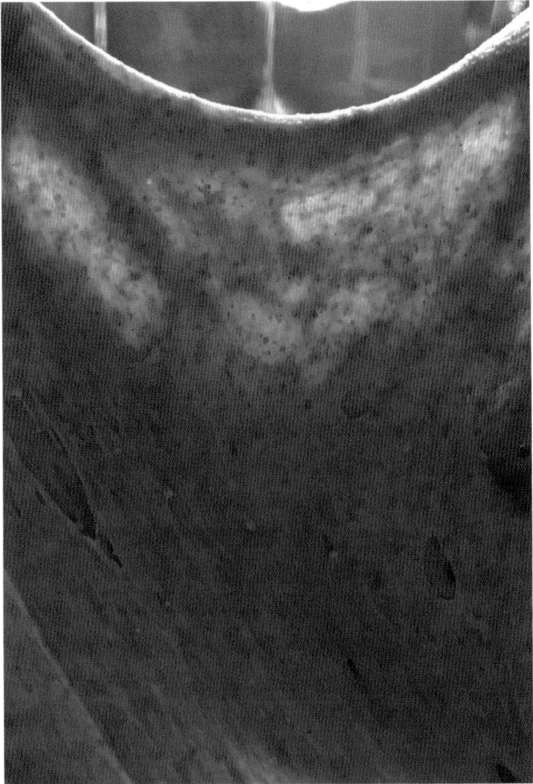
A windowpane in whole grain flour dough

Poke Test

Eventually your dough will have risen and you'll want to know if it's ready. "Ready" means it has enough gas and it has a strong gluten network to hold the gas in the dough through baking. There is also a stage way past ready called over-fermented in which the yeast has eaten all the food in the dough and will no longer be able to make new gas, and the bacteria in the dough have decided to eat the gluten network. You can see this would be bad—no gas, no gluten network, no puffy bread. The poke test can help you figure out if your dough is ready and if it is in danger of becoming over-fermented anytime soon. I find this test is also useful in assessing whether a loaf is ready to bake.

A caveat: The poke test is not as useful in whole grain doughs and should never be used on refrigerated dough, which is too cold and stiff to spring back.

For a poke test, poke a wet finger half an inch into the dough for a second, then remove.

- Does the hole fill right up? It's not ready yet.
- Does the hole start to fill back up over the next few seconds, but not completely? It's probably ready.
- Does the hole stay there just as you left it or does it take a long time to start filling back in? Your dough is possibly ready or past ready. (This doesn't mean it's over-fermented, but you should move quickly through the rest of the recipe to keep it from becoming over-fermented.)

Smell Test

Wet flour smells like just that. Once yeast has fermented the flour, it smells different, more like dough, a good bakery, yeast, or bread. In sourdough, you can also start to smell the tartness of the acids as the dough ferments. I find this test most useful in assessing the bulk fermentation.

For the smell test, sniff your dough just after mixing, then pay attention to the smell each time you check it. You'll be able to tell from the smell about where it is in the process, but this test is not exact.

- Wet flour or gluey smell—definitely not ready
- Doughy smell—getting there
- Yeast, bread, or bakery smell—possibly ready or getting close
- Slightly acidic smell—possibly ready or getting close
- Strong acid smell—possibly ready or past ready
- Very strong vinegar smell—probably past ready

Jiggle Test

This is my favorite ready test for assessing the bulk fermentation, because when my bowl of dough jiggles, the bread is always good. Plus it's easy. When the dough is full of enough gas and is held in place by a strong and stretchy gluten network, it will jiggle like jelly or a water balloon. The jiggle test is not useful for lower hydration doughs or cold dough. Note that in some recipes (such as those with a lot of whole grains or a cold retard), we end the bulk fermentation before peak jiggliness to avoid over-fermentation.

If your dough has gone past jiggly, it may become liquid or soupy and remind you of the slime toys kids play with. This is over-fermented dough and the gluten network is degrading. See the Troubleshooting section (page 276) for help.

To perform the jiggle test, simply shake the bowl a little and see if the dough can jiggle (meaning it keeps moving after the bowl is still). As soon as it can jiggle, it's plenty ready. If you think it's almost jiggling, it's probably very close. You should check again in 5 to 15 minutes so it doesn't over-ferment. If the dough can slosh around merrily with lots of large bubbles, it is really ready and you should preheat your oven and move efficiently through the rest of the recipe to keep it from becoming over-fermented.

Shaping the Dough into One or More Boules or Batards (10 to 20 minutes)

Turn Out and Divide

Once the dough is ready, it's time to shape it into loaves. In commercial-yeast baking, you would now punch down the dough. *Do not* punch down your lovely, lofty sourdough. We want to shape the loaf in a way that preserves as much as possible the gluten and bubble network that has formed. We do need to deflate the dough a little to form the loaf and to give it a chance to rise again. If you want an open crumb, you should try to keep some of the puffiness of the dough during the shaping.

You can mist your counter with a light amount of water or sprinkle it with a light amount of flour for this step. I suggest which method I find easiest for each recipe. The dough will be very sticky, but we don't want to add in too much flour or water at this point because it isn't helpful and it upsets the hydration level in the recipe. If you touch the dough only when your hands or implement are wet, floured, or oiled, then it won't be a problem that the dough is sticky.

To turn out and divide the dough, prepare the counter with a little dampness or a little flour. Pour out the dough, using a wet silicone spatula to scrape all the dough out of the bowl in one piece, as much as possible. It will look like a sticky, bubbly mess. If you are making more than one loaf, use a bench knife to divide the dough. Use a scale if you want the loaves to be the exact same size.

Pre-shape

To get the dough into a loaf, we stretch it into the shape we want. If we do this too fast, the dough will tear, kind of like muscles. So we do it in stages, with a rest in between for the gluten to relax and be ready for the next stretch. The process of pre-shaping and shaping also moves the dough around, allowing yeast to find new pockets of flour to eat, causing the loaf to rise.

After you pre-shape the dough it will have two very different sides. The seam side, which is messy and sticky, and the top side, which is smooth and less sticky and will end up being the top crust of the loaf. You will flip your pre-shaped dough seam side down and lightly sprinkle flour on the smooth top. You don't flour the seam side because you want that side to be sticky—the stickiness allows you to close the seam when you shape the loaf.

During bench rest, the dough will spread into a flattened mound. How flat will depend on the flours used and the hydration level as well as the quality of the gluten network. But generally, the dough should take at least 5 to 10 minutes to relax and should still be a mound shape with a rounded edge where it meets the counter. If the ball flattens out dramatically or rapidly, or if the edge slopes down to the counter, a second pre-shaping can impart more strength to the dough.

Sliding a bench knife under the left side

Folding it to the center

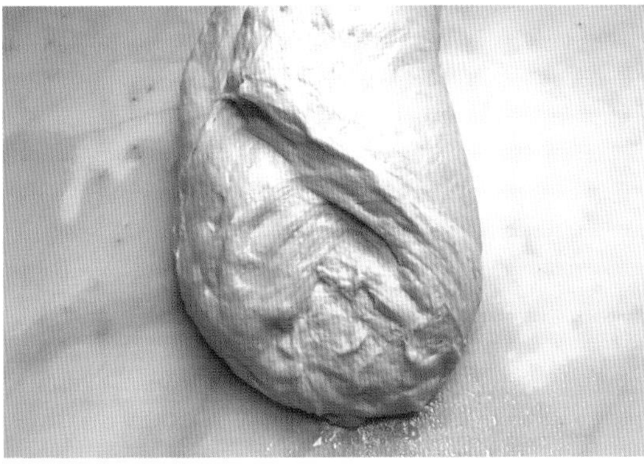

After folding the right side to the center

To pre-shape, fold the dough in thirds (like a business letter), twice, just like you did for a stretch and fold. In this case, we are using a lightly floured or damp bench knife.

1. Decisively scrape the bench knife under the left side of the dough. Using your wet or floured hand to guide the dough, stretch it out a little and then fold it onto the center of the dough and hold it there to adhere. This can seem hard because it's like trying to fold a big, floppy water balloon, but don't be afraid to stretch the dough up and press it over. As long as you don't see tearing, it's all good.
2. Repeat on the right side, the top, and the bottom.
3. Use the bench knife to flip the ball of dough so the seam side is down.
4. Using wet or floured hands, cup the ball of dough and rotate it like it's the lid of a giant jar. The bottom will stick to the counter and the skin on top will stretch and tighten, causing the dough to rise into a sphere. Watch to be sure the skin is not overstretched to the point of tearing.
5. Rest the dough on the counter. If you used water while pre-shaping, dry off the counter and the bench scraper while the dough rests.

Shape the Loaf

Once the dough has relaxed back into a flatter shape again, it's ready to shape into a boule (round) or a batard (oblong). Shaping does more than just cause the dough to be round or oblong. It causes a skin to form on the outside of the dough that acts as a support structure during rising and baking, kind of like how the walls of a building hold up the floors and the roof. We want to form the skin, but not tear it, so our loaf is tall and stable, and we need to seal the skin at the bottom of the loaf so it stays taut. Although there is no baguette recipe in this cookbook, I include instructions for shaping a baguette in this section for your information. If you want to try a baguette, choose a recipe with at least 50 percent white flour and proof it on a parchment paper–lined baking sheet. The Simple Boule (page 91), Crusty Boule (page 95), or Faithful Batard (page 117) recipes would make good baguette dough.

When it's time to shape the loaf, sprinkle small amounts of flour on your hands and counter, then flip the dough top side down so the sticky side is up during shaping. Remember—hot potato! Don't touch it for long. After shaping, the top side should be more satiny than sticky. Try not to totally deflate the dough or to add a lot of flour during this step.

Shape a Boule

1. Flour the top of the dough and flip it onto a *lightly* floured counter.
2. Using lightly floured hands, pull up one edge of the dough and press it into the far side for a couple of seconds to adhere.
3. Continue around the ball once or twice, letting the dough tell you when to stop. (When it resists pulling or begins to tear as you are pulling, it is time to stop.) Some doughs, such as high-hydration whole wheat, can only handle 4 of these manipulations, while a very stretchy dough with no whole grains may be able to withstand 8 or 10. Brush off any excess flour from the dough as you go so it won't end up as a tough layer in your loaf.
4. Pinch together the seam and hold until it sticks closed. This is harder to do if you used a lot of dry flour.
5. Flip the ball onto an unfloured area on the counter.
6. For a taut skin on a boule, you must tighten it. Here is the method I like to use: Pull the ball toward you with the bench knife or your cupped hands while it is sticking to the counter. Watch the skin on top of the ball stretch as the bottom of the dough sticks a little to the counter and you pull the ball toward you. Turn the ball and repeat around the ball until it stands round and tall, but immediately stop if there's any tearing. The goal is that the skin on top of the dough seems taut but has not begun to tear. If you want to skip this step, your loaf will just be a bit flatter and will spread out more in the oven.
7. Allow the ball to sit for a minute so that the bottom seals closed.

Pulling up one edge of the dough

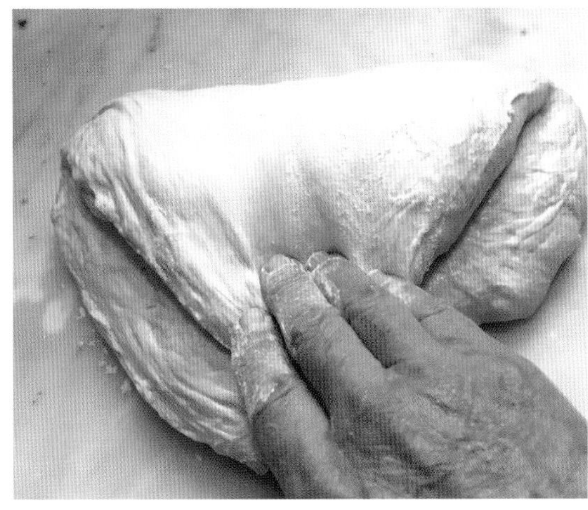
Pressing it to the far side

Pulling up the next edge of the dough

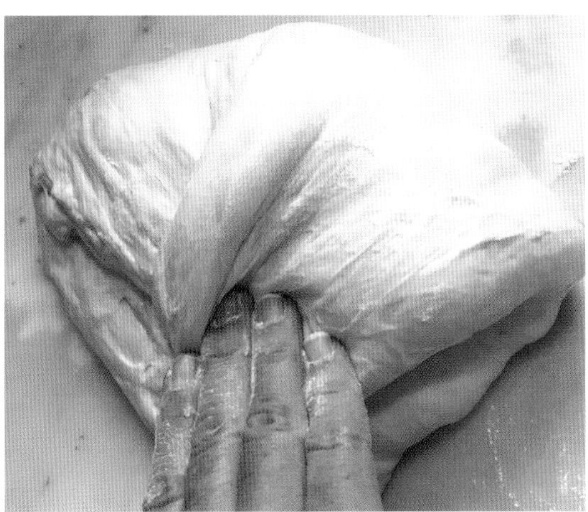
Pressing it to the far side

Tightening: Sliding the bench knife under

The skin has stretched

Shape a Batard

1. Flour the top of the dough and flip it onto a *lightly* floured counter.
2. Lift the right side of the dough and fold over the center and hold it there a second until it adheres. Repeat with the left side.
3. Starting at the top, roll the dough down over itself and press with the sides of your thumbs at where it meets itself to adhere, gently pushing the roll away to stretch the skin. Then continue rolling and press the next seam and the next until the roll is finished.
4. Pinch together the seam and the ends to close them securely.
5. If desired, tighten the skin by using a bench knife on the long side of the loaf to pull it along the counter, then repeat on the other long side.
6. Allow the batard to rest seam side down for a minute so that the bottom seals closed.

Sides folded in and ready to begin rolling

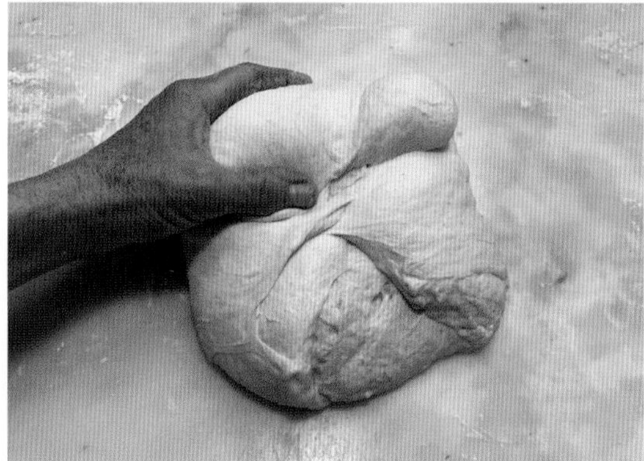
Starting to roll

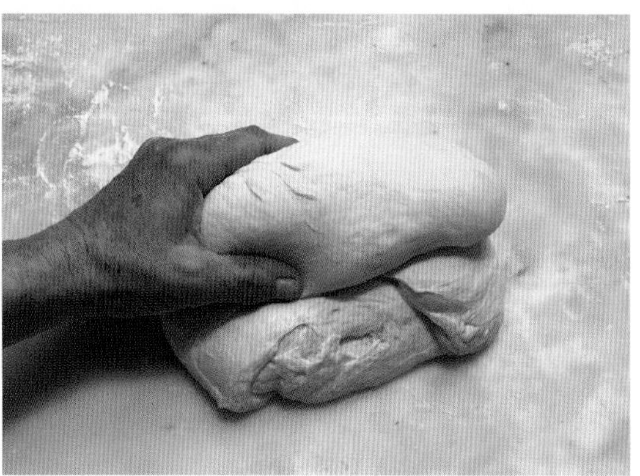
Continuing to roll

Finished rolling

Shape a Baguette

1. Flour the top of the dough and flip it onto a *lightly* floured counter.
2. Lift the top side of the dough and fold it to the center and press it there a second until it adheres. Repeat with the bottom side. The area where you press will become the hinge in the next step.
3. Now fold it in half, bringing the top side onto the bottom. It may be necessary to press some gas out along the centerline so it can fold.
4. Pinch all along the seam and the ends to close it tightly.
5. Roll the loaf seam side down and rest a few minutes to seal the seam and to relax the dough a bit.
6. Stretch the baguette gently and evenly until it is 14 inches (36 centimeters) long.

Lifting the top side to the center

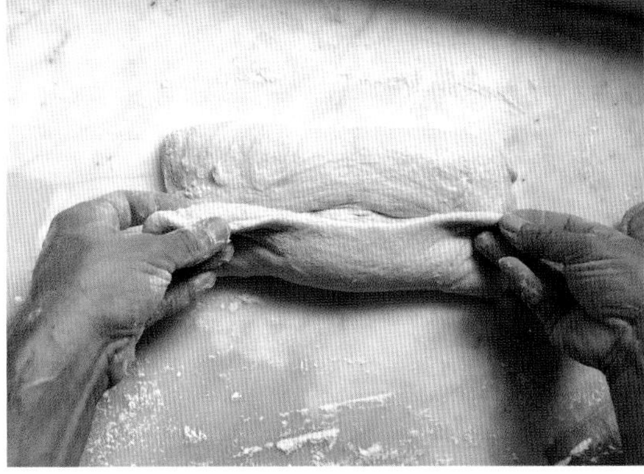

Repeating with the bottom side

Folding in half

Ready to pinch the seams closed

Proofing the Loaf (30 minutes to 24 hours)

Prepare a Proofing Basket

Once the loaf is shaped, it needs to proof before baking. This step gives the dough a chance to rise again and develop more flavor. For most artisan loaves, a support called a proofing basket is used to contain the dough while it proofs so the resulting loaf can be tall rather than flat and wide. A proofing basket is a container that supports the dough during its final rise and is removed just before baking the loaf. It should be 7 to 9 inches in diameter for a boule, or 4 to 5 inches wide by 9 to 10 inches long for a batard. It must be an inch or two smaller than your covered baker or surface, and it must have upright sides to support the dough. It is best if the basket can allow some air circulation. No matter the type of proofing basket, a cloth liner can be used. The cloth must be tightly woven, smooth, and preferably made from linen or cotton. Do not use terrycloth, which sticks mercilessly to dough. A proofing basket can be any of the following:

- A colander lined with a well-floured, smooth cloth such as a tea towel for a boule.
- A bowl lined with two well-floured, smooth cloths such as tea towels for a boule. The extra towel helps the dough to breathe.
- An oval or round basket lined with a floured, smooth cloth for a batard or a boule.
- A sling or hammock fashioned from a floured cloth to hold a batard.
- A special basket designed for bread proofing, such as a banneton, to hold a boule or a batard.

Note that some of these special proofing baskets must be seasoned before use according to the manufacturer's recommendations. Alternatively, they can be lined with a floured, smooth cloth.

To prepare the proofing basket, line it with a cloth if you are using one (recommended for very wet dough or overnight retards). Using a sifter or mesh strainer, sprinkle flour (preferably rice flour) evenly around the entire basket, including up the sides where the dough will rise and want to stick.

Proof the Loaf

During proofing, we need to prevent the dough from drying out and getting a tough skin, so it must be covered in a way that keeps it moist, accommodates any rising that may happen, and keeps anything the dough could stick to off the surface of the loaf. Sprinkling a little flour, then covering the dough with parchment paper keeps it from sticking to the cover or bag used to keep the dough moist.

The loaf should rest at the temperature called for, usually room temperature (60–80°F; 16–27°C). It will rise faster at the higher end of the temperature range. I refer to this as a warm proof. The range of times given in the recipe alert you that the baker needs to check on the loaf, using a poke test or jiggle test to determine when to bake. For the best flavor, loaf proofing can be done in the refrigerator. This is called a retard, because the low temperature delays the rising. There is a long window for the timing of a retard. You will find that most rising is complete within the first three hours. After that, flavors will continue to develop, but the loaf is safe from over-fermentation until the end of the time range, so baking can happen whenever it's convenient. The usual "ready" tests don't work on cold dough, so you should adhere instead to the proofing time range in the recipe.

THERE ARE SEVERAL BENEFITS TO A RETARD:

- Bake on your own schedule, taking the dough out when you want to bake.
- Cold dough is easier to score beautifully or with intricate patterns.
- Great oven spring, that lovely burst of expansion that takes place after the loaf goes into the oven, is attained by placing cold dough in the hot oven.
- More flavor development is achieved because of the longer overall fermentation time.
- It's easier to avoid accidental over-proofing, which is especially nice for doughs with high percentages of whole grains or if you live in a hot climate.

A batard in its lined banneton, ready to be covered in parchment paper and placed in a plastic bag to keep it from drying out during proofing.

THERE ARE SOME CONSIDERATIONS FOR RETARDING A LOAF:

- During the long proofing time, dough can stick to the proofing basket, but using plenty of rice flour and a cloth liner prevents sticking.
- The loaf can develop more acetic acid tang, which may not be desired.
- The dough can dry out, forming a tough skin, but covering it well or using a plastic bag to enclose the loaf prevents this.
- The loaf can still over-proof, especially if the dough already experienced a cold retard during bulk fermentation, the recipe has a lot of whole or sprouted grains, or there are signs that the bulk fermentation went too long. If you know you want to retard the loaf, be sure to end the bulk fermentation on the early side.
- You have to wait longer before you have bread and must plan accordingly.

To proof the loaf, flour the top and sides of your loaf. Use your bench knife to lift the loaf from the counter and to flip it onto your floured hand. Place the dough seam side up into the prepared proofing basket. Sprinkle some flour on the dough and cover it with parchment paper. Cover the basket with a lid that leaves plenty of room for the dough to rise or place the entire basket into a plastic bag and tie it closed or place a big mixing bowl over the entire basket. Decide if you are using a warm proof and will be baking soon, or if you are retarding it and baking later, and follow the next instructions accordingly.

Warm proof: Rest the dough at the proofing temperature for your recipe for the allotted time. Use the poke test or jiggle test to see if it's ready. The proofed loaf should seem like a water balloon. It is better to under-proof the loaf (because oven spring can still raise the loaf), than to over-proof it (it won't rise or may even fall in the oven).

Retard: If you retard your loaf in the refrigerator, keep it refrigerated until just before placing it in the hot oven. The poke test and jiggle test don't really work on refrigerated dough; you must use the time range called for in the recipe. The longer end of the time range is best if you want more flavor in your dough or if you suspect you under-proofed during the bulk fermentation. Shorter times are better if you don't want much sourness. If you are concerned that the dough over-proofed in the bulk fermentation, you shouldn't retard the dough but instead bake it as soon as your oven is hot.

Baking (1½ to 2 hours)

Artisan breads are typically baked at very high temperatures to get that distinctive rich chestnut brown crispy crust. Most of the recipes in this book call for 475°F or 500°F (246°C or 260°C). If your oven doesn't go that high, you just need to bake a little longer. It can take extra time for home ovens to get up to these temperatures, so plan accordingly for a long preheat. Temperatures in a home oven plummet whenever the door is opened, so it is important to resist the temptation to keep opening the oven "to check."

The baking of bread has two phases. During the first phase, usually 15 to 25 minutes, the bread rises, creating oven spring and a lofty crumb. Just before baking, we score (cut) the skin of the loaf to allow expansion and to control where the bread expands. We provide steam to keep the crust stretchy long enough that the dough can expand fully. Professional ovens have steam jets, or they are tightly sealed after loading the bread to trap steam coming from the dough. The steam also cooks the crust, so the longer the steam phase, the thicker the crust will be on the loaf. During the second phase, usually 15 to 30 minutes, the crust browns and the interior bakes. For the crust to brown nicely, we need to eliminate the steam.

Prepare Your Baking and Steam Setup

Although it is hard to match professional ovens, there are many possible ways to create steam and to bake very good artisan bread using your home oven. Some home ovens do have a steam option, but the rest of us must rely on other methods. Following are three steaming methods. For any of these methods, you can mist the loaf generously just after scoring to provide additional steam, if desired. In order to keep the steam trapped, it is important not to open the oven or to lift the cover until it's time to end the steam phase.

STEAM SETUP 1: Bake in a covered baker such as a Dutch oven, a casserole dish, a ceramic cloche, a graniteware roaster, or a stockpot that is at least 3 or 4 quarts (3 or 4 liters). At the end of the steam phase, the cover is removed to release the steam. This method works by trapping the steam released from the dough inside the pot until the cover is removed. This method has the highest probability of yielding success for a beginner. Beware that some covered bakers or their handles are not made to withstand a 500°F oven (260°C). Covered bakers can be preheated with the oven if they are metal, otherwise follow manufacturer's guidelines. Never preheat glass. Hot covered bakers present a real danger of burning you when you go to put in the loaf. They can also crack the oven's window glass if placed on the oven door. I notice no difference in bread that's baked in preheated versus room temperature covered bakers, provided the steam phase is extended by 10 to 15 extra minutes in the case of the room temperature baker (presumably to heat up the baker). Unless you have an oval covered baker, this method can only be used for boules.

STEAM SETUP 2: Bake on a baking stone or sheet that's covered with an upside-down, room temperature metal mixing bowl, stockpot, or deep roasting pan. The cover over the loaf traps the steam released from the dough until the cover is removed at the end of the

My favorite baking method is to bake on a stone, covering the loaf with a giant mixing bowl for a boule or an oval roasting pan for a batard.

steam phase. This is my favorite method! It's easy and, provided you have a nice wide cover, it can accommodate batards and rolls as well as boules. It is important that there are no gaps at the edges of the cover where the steam can escape.

STEAM SETUP 3: Bake on a baking stone or sheet with a steam pan. Place a heavy, cast-iron or roasting pan on the bottom of your oven during preheat. Once the loaf is in the oven, water is added to the hot pan to create steam and the oven door is quickly closed to trap it. The water boils off after 15 to 20 minutes, naturally ending the steam phase, and then the loaf will brown. The steam can really billow up and burn you when adding water, so some bakers like to add ice cubes. However, ice will bring down the oven temperature. Some ovens provide better results with this method than others that may leak or vent the steam too quickly. One clear advantage to this method is that you can see the bread during the steam phase of the baking process, provided you have a window in your oven door. It's fascinating to watch oven spring!

You have to experiment with your own baking and steam setup to see which gives you the bread you prefer.

To prepare for baking, preheat your oven 30 to 60 minutes before you plan to bake. Adjust the racks; generally the second lowest rack is a good bet. Make sure your baking setup fits. Baking stones, cloches, and cast-iron pots usually require 45 minutes to an hour to preheat—place them in the oven at the start of the preheat. Use an oven thermometer to verify the temperature. Gather your tools, which may include the following.

- Oven thermometer
- Peel
- Pastry brush
- Lame (for scoring the bread)
- Water mister
- Covered baker
- Hot mitts
- Cover
- Timer
- Pitcher of water
- Instant-read thermometer

If your loaf is ready but your oven isn't hot enough, put your loaf in the refrigerator while you wait for the oven to come up to temperature.

Turn Out the Loaf

Wait until your oven is hot and your setup is all prepared.

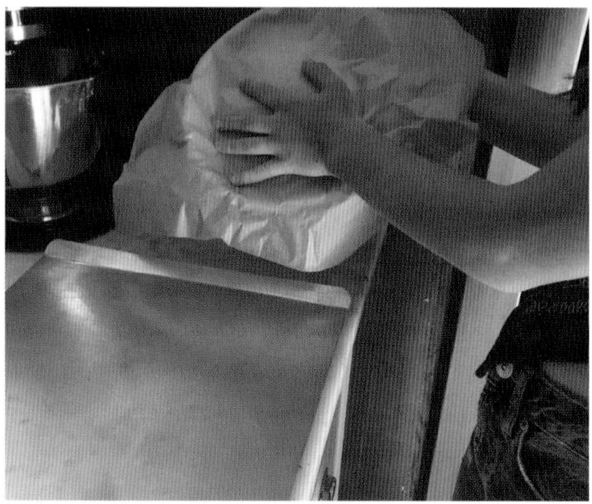

Guiding the loaf

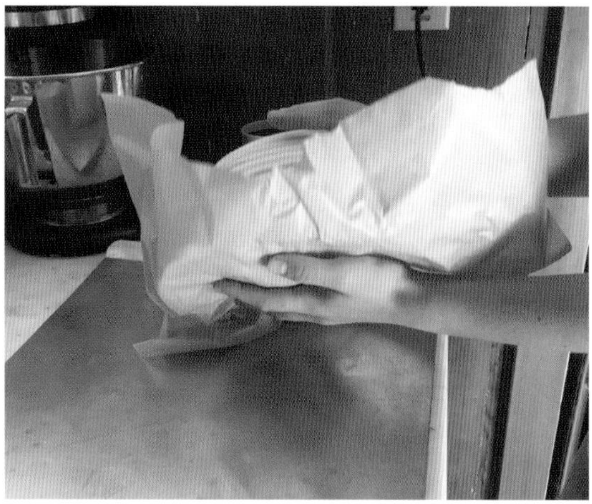

Tipping the proofing basket

Lifting the proofing basket

Score and Mist the Loaf

The goal of scoring is to cut through and just under the skin of the loaf without cutting down into the dough. The dough will be able to expand and stretch at the cut, but you don't want to compromise the structure of the loaf by cutting too deep (more than half an inch; 1 centimeter) because then the loaf would spread rather than rise. If you hold the blade at a low 30- to 45-degree angle to the surface of the dough, you can get a good cut under

What a cute loaf!

To turn out the loaf, hold the proofing basket in your dominant hand and place the other hand gently on top of the parchment paper covering the loaf. Tip the basket over the counter or peel and guide the loaf with your hand onto the surface. Lift the proofing basket away. The loaf will seem to flatten and spread. Brush off excess flour gently with a pastry brush (optional).

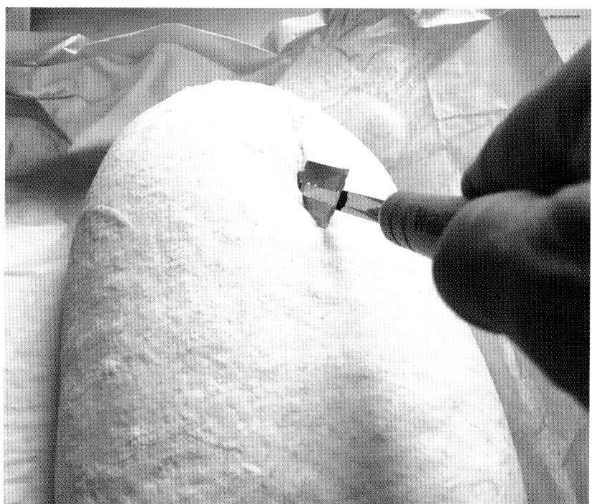

Cutting under the skin at a low angle

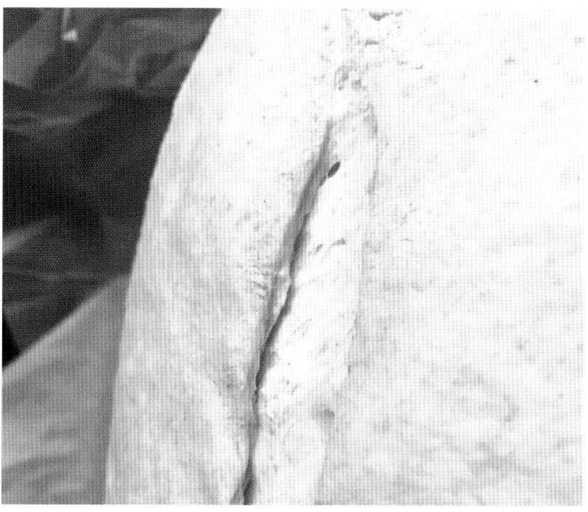

The skin has been separated and is lifting away from the dough, but the dough itself is not deeply cut.

Crispy ears

the skin without going too deep. A single cut along the length of the loaf is always a safe bet, but there are lots of patterns that can be used. A shallow cut will not go all the way through the skin but will stretch a little in the oven, so shallow cuts are mainly decorative. You can use a lame, a razor blade, a kitchen knife, or kitchen scissors, but whatever you use must be sharp. It is easier to score cold dough because it is stiffer.

A loaf scored as described here may develop a crispy ridge of crust that rises and separates from the loaf during the steam phase when the loaf expands rapidly. This is affectionately called the ear, and it is sought after by some bakers. To achieve an ear, the loaf proofing and bulk fermentation must not go overly long, and the loaf must be well shaped and tightened, followed by skillful scoring. However, a loaf does not need an ear to be excellent—if you like your bread, don't worry about the ear.

Doughs made with mostly or completely whole grain or very high hydration sometimes rise better with little or no scoring (I will state this in the recipe if it applies). When a strong gluten network is not present in the dough, loaves are likely to spread at

Two loaves from the same bowl of dough: the one on the left was misted before baking.

the cut rather than rise. The same goes for very over-proofed doughs with a gluten network that is degrading.

After scoring, you can mist the loaf and its parchment paper generously with water using a spray bottle. This optional step provides extra steam to help with oven spring and gives a shiny crust. It also helps soften the parchment paper so it doesn't interfere with sealing the cover or leave wrinkle prints in your loaf if you are using a covered baker.

To score the loaf, just before baking, cut about ½ inch (1 centimeter) into the loaf in one or more swift, decisive cuts. You should be able to see the bubbly dough under the skin of the loaf. Mist the loaf and its parchment paper generously using a spray bottle (optional).

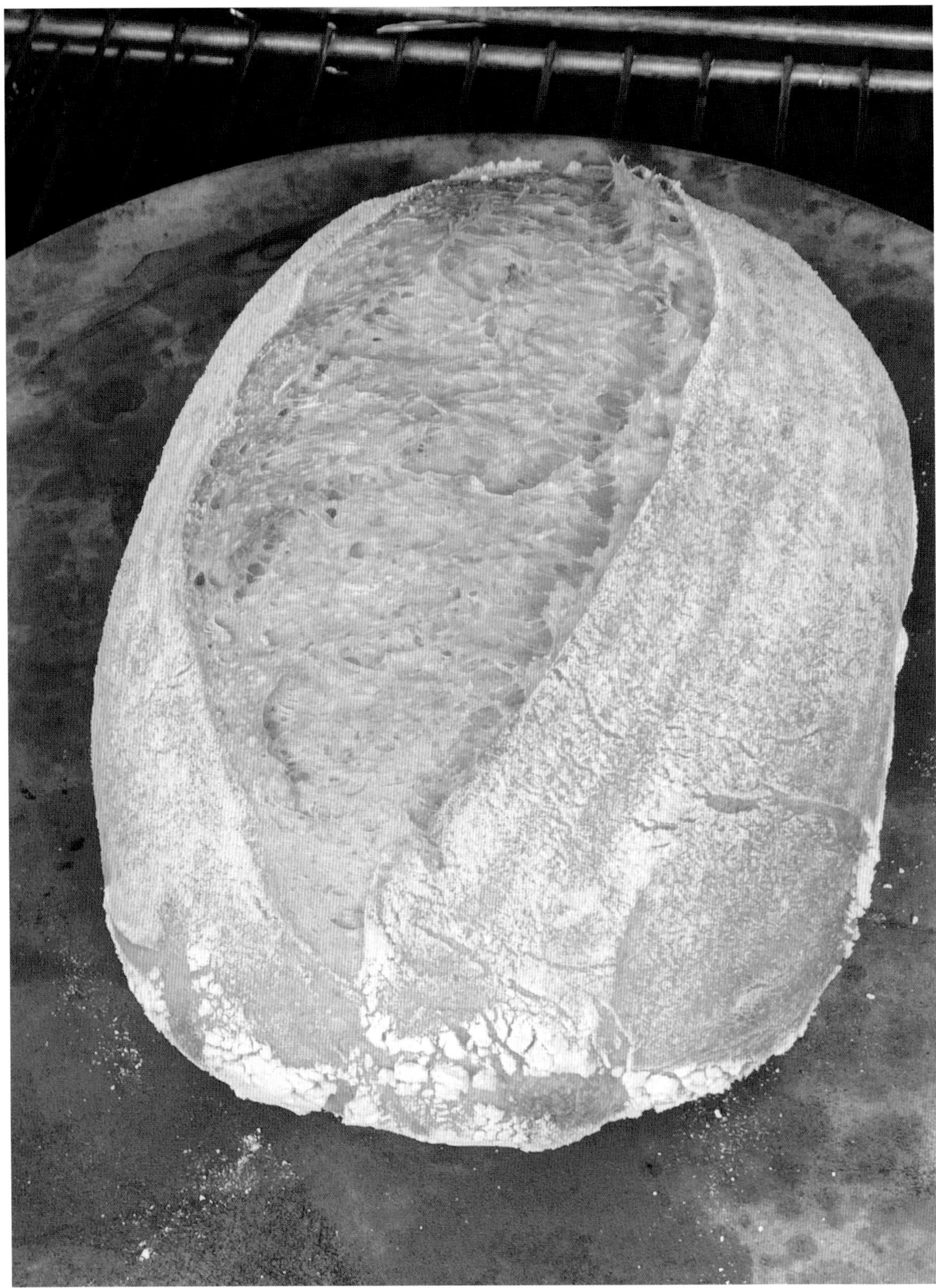

A loaf at the end of the steam phase

Bake

Baking is when the dough cooks and the crust browns. For a crusty, flavor-packed artisan loaf, you must give enough time for both of these processes to complete. It is common for beginners to underbake the loaf, thinking it is done when a honey gold crust glows at them from the oven. And who is to say a lightly browned crust is not good? You should bake the bread you like to eat, but the interior needs to be fully baked. Once the internal temperature is at 207°F (97°C), you can be sure the crumb is completely baked. At this point the bread will give a hollow, resonant drumlike sound when rapped on the bottom, rather than the dull thud of an underbaked loaf. Before then, the crumb may be gummy and undercooked in a sourdough. Some people do prefer the doughy, yeasty crumb of sourdough bread baked to only 195°F (91°C), but this is the minimum temperature that will yield bread rather than dough in the center of the loaf.

Many of us are after a crust that is crispy and loaded with flavor. Even after the crumb is finished cooking, continued baking will cause the crust to brown and flavor with caramelized sugars and products of the Maillard reaction. Steam from the dough inside carries flavor compounds and concentrates them in the crust as it passes through. If you take the bread out too soon, you will miss out on the flavorful crust these processes provide. It is very hard to burn bread if you are checking on it every 2 to 5 minutes once the internal temperature reaches 207°F (97°C), so try to resist taking the bread out too soon. The crumb is so moist due to the high hydration and sourdough fermentation that there is no reason to worry about a dry interior at the temperatures used in these recipes.

To bake the loaf, choose the instructions for your chosen baking setup.

Covered baker: Transfer the loaf and parchment paper into the enclosed baker by lifting it by the parchment corners. I like to mist the inside of the lid with water before placing it over the loaf for extra steam. Cover and place in the oven, leaving the cover undisturbed for the duration of the steam phase. Wearing an oven mitt, remove the cover to end the steam phase, being careful of the hot steam (remember to extend the steam phase by 10 to 15 minutes if the baker was not preheated). I get a more even crust color by removing the bread from the baker and placing the loaf directly on the oven rack to finish baking.

Baking stone or sheet with a cover: Use a peel to slide the loaf and its parchment paper onto the center of the hot stone, or place the baking sheet with the loaf in the oven. Mist the inside of the cover with water before placing it over the loaf for extra steam. Make sure there are no gaps for steam to escape around the cover, then leave the cover undisturbed for the duration of the steam phase. Wearing an oven mitt and perhaps employing tongs, remove the cover, to end the steam phase, being careful of the hot steam.

Baking stone or sheet with a steam pan: Use a peel to slide the loaf onto the center of the hot stone, or place the baking sheet with the loaf in the oven. Once the loaf is in the oven, wearing an oven mitt, pour 1 to 2 cups (250 to 500 milliliters) of water the hot pan to create steam, quickly close the oven door, and do not open it until the end of the steam phase. The water boils off after a few minutes and is vented from most ovens, which will end the steam phase automatically. However, the oven door can be opened at this point to be sure all the steam is vented. The steam pan should be removed at this time if it still has water in it.

Once the steam phase is over, continue baking until:

- The interior of the loaf is 207°F (97°C). Then begin checking the crust color every 2 to 5 minutes, rotating or moving the loaf as needed for even browning. The parchment paper can be removed if desired.
- The loaf sounds hollow when you tap on the bottom.
- The crust looks good to you, generally chestnut brown to deep cocoa brown gives the most flavor. When in doubt, bake longer.

If the loaf is becoming too brown before the inside is baked, turn down the oven temperature by 25 or 50°F (14 or 28°C).

Anything that allows air to flow around the crust can substitute for a wire rack.

Cure the Crust

Steam from inside the loaf will escape through the crust and soften it as it cools. To keep the crust crisp, cure the crust in the cooling oven.

To cure the crust, just after baking, turn off the oven, prop the oven door ajar, and place the loaf directly on the wire rack in the cooling oven for 5 to 10 minutes.

Cool the Loaf

The inside of the loaf is still cooking after it is removed from the oven. The loaf will be best if it is allowed to cool slowly in a warm spot with good air circulation, such as a wire rack on a counter near the cooling oven. Cutting it too soon will result in a gummy interior. But . . . this isn't much of a problem if it all gets eaten before it cools. Still, try to give it at least an hour to cool for the best texture.

To cool the loaf, place it on a wire rack for at least an hour or two before cutting or storing it.

Part One
Introduction
Part Two
How to Make
Artisan Sourdough

Part Three
The Artisan Bread Recipes

Part Four
Traditional Breads
Using Your
Sourdough Starter
Part Five
Appendix

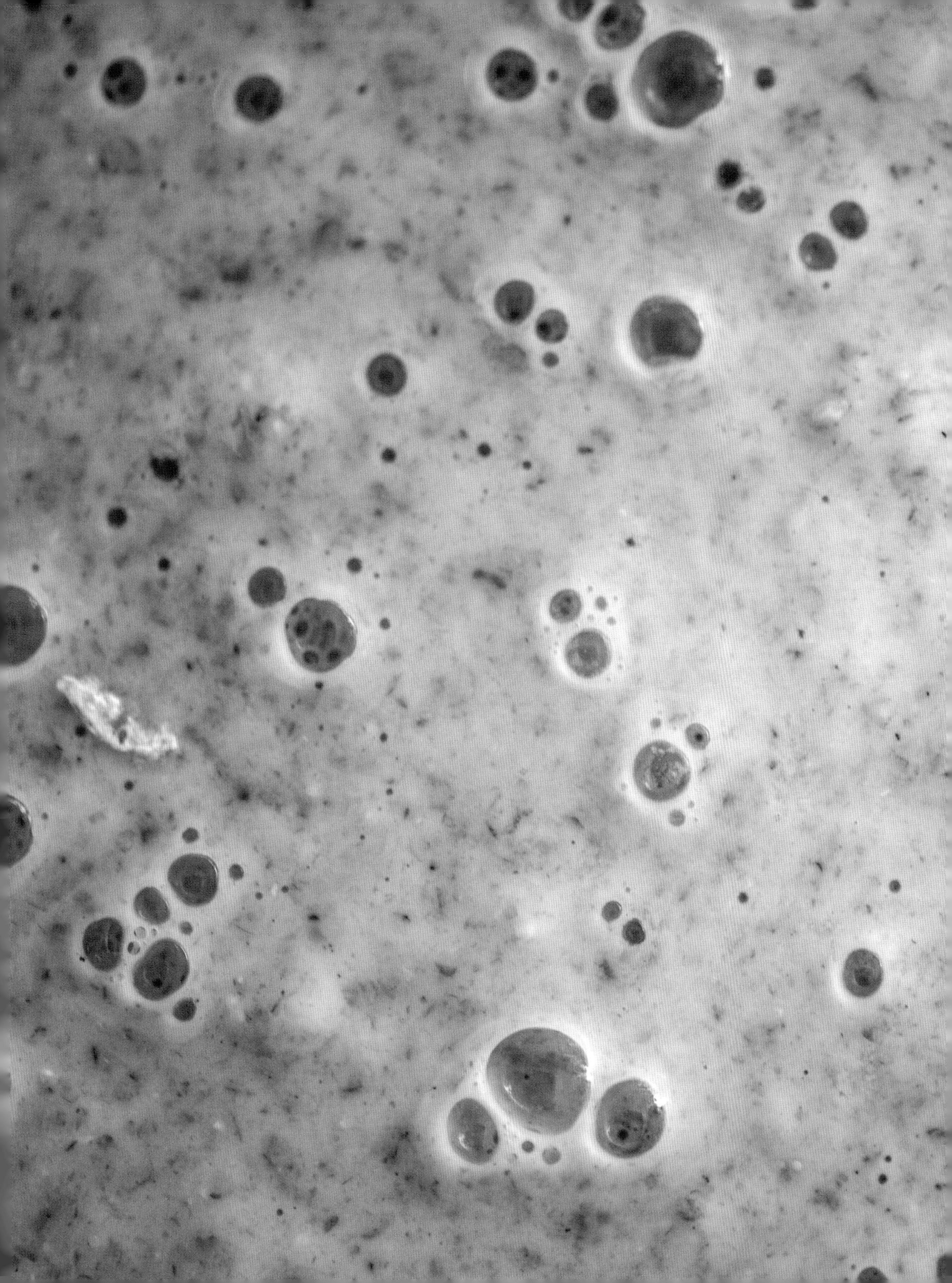

CHAPTER 6

Essential Recipes: Developing Your Sourdough Intuition

Why do some sourdough bakers consistently turn out outstanding loaves while others suffer sporadic hits and misses? A powerful way to become proficient at sourdough bread baking is through experience coupled with record keeping and reflection. In this chapter, I walk you through three recipes that bake up delicious loaves of sensational-looking artisanal sourdough bread. The recipes in this chapter include very detailed instructions to ensure your success. The recipes in the later chapters are more condensed for your convenience, as they assume you've gotten a feel for the commonly used techniques.

Bake Three Loaves

To boost your sourdough-baking prowess, select one of these recipes and bake it several times over the course of a few weeks. Three bakes is generally the magic number. Start by baking it as closely as you possibly can to the recipe as written. (For those of you who like to improvise—I hear you and I get it, but save that creative renegade spirit for just a little later. We are building the foundation here!) Take notes using the Sourdough Baking Worksheet (page 290) and spend a few moments thinking about what worked, what didn't, and what you'd change the next time. Then bake it again, making adjustments to suit your starter or equipment, weather, schedule, and so forth. Then bake it a third time. By then you will probably be getting pretty comfortable with the rhythm, the terminology, and the techniques. Perhaps you are even gaining an intuitive sense of whether your levain or your gluten development or your proofed loaves are ready for the next step. It is this sourdough intuition that separates the celebrated baker from the terminally mediocre or inconsistent one.

Sourdough Intuition + Hands-On Skills + Knowledge = Amazing Bread

In the process of baking your three loaves, you will also gain the hands-on skills you need to be a proficient baker. Handling high-hydration dough, shaping a good boule or batard, and tightening its skin are skills that you really have to learn by doing a bunch of times. If you want to up your savoir faire in this area quickly, I suggest making the recipe for Octo Rolls (page 103) as your third or fourth bake. You get to pre-shape, shape, tighten, and score eight times in one recipe. You can try different tools and techniques for handling the dough, make some into mini-boules and some into mini-batards, or experiment with different ways of tightening the skin or different scoring patterns to see which works best for you in your kitchen. You can also divide the dough into two mini-loaves in any of these Essential Recipes and compare different techniques or baking and steam setups.

As you bake, refer back to Chapter 5 as needed to enhance your understanding of the science behind what's happening in the bowl of dough. If you have any trouble at all as you go through the recipe, consult the Troubleshooting section (page 276). It is there to help you get out of binds, avoid the problem the next time, and to offer rescue options for recipes gone awry.

The recipes in this chapter all begin with a levain, which is a special large batch of starter that is made up fresh the day before mixing the dough. Using a levain ensures that you are baking with lively, happy yeast. Levain recipes also provide ample opportunity to handle and observe the dough as it goes through its bulk fermentation, which is an excellent way to gain a lot of knowledge and experience. I've included estimated times for each step in the recipes to help you plan out a successful baking adventure.

Ready? Let's bake.

Simple Boule

Sometimes the best things are simple. This boule showcases the pure, unadulterated bliss that artisanal sourdough delivers. It's a great recipe to begin with, as it calls for only white flour with a medium hydration level. Plus, it's a great loaf to eat! For the best sourdough flavor and a blistered crust, retard the loaf overnight. Baking the same day will give a more delicate flavor to the bread.

MAKES ONE LOAF

TOTAL FLOUR: 450 GRAMS (1 POUND)

PRE-FERMENTED FLOUR: 17%

HYDRATION: 68%

WHOLE GRAIN: 0%

DOUGH: 765 GRAMS (1.7 POUNDS)

DAY 1

Build the Levain (10 minutes)

1. In a small bowl or pint jar, mix together:

30 grams starter that floats (2 tablespoons)

60 grams water (¼ cup)

60 grams all-purpose flour (½ cup)

2. Cover and let it rest at 60–70°F (16–21°C) for 8 to 12 hours until it is risen and bubbly.

DAY 2

Test the Levain (a couple of minutes)

It should be risen two to three times the original height and bubbly, but not already deflated (a little loss of height is okay). If you drop a spoonful into water, it should float.

Mix the Dough (10 minutes)

1. Measure into a large heavy bowl:

375 grams bread flour (3⅛ cups)

2. Whisk together in the levain bowl or jar until uniform:

231 grams water (scant 1 cup) at 90°F (32°C)

150 grams levain (all)

3. Combine completely with a silicone spatula until all the flour is wet and it forms a ball. This usually takes 2 to 3 minutes. Toward the end, you can use a damp hand to finish mixing, if needed. Do not add additional flour or water.

continued

Autolyse (30 minutes)

1. Cover the bowl with a lid or plate and let the dough rest for 30 minutes at 74–80°F (23–27°C), or until it is relaxed and soft and gives a nice windowpane.

2. Add:

9 grams salt (1½ teaspoons)

3. Sprinkle the salt evenly on the surface of the dough and poke and pinch it deep into the dough with your wet fingers.

4. In the bowl, thoroughly mix the salt into the dough with the silicone spatula or a wet hand.

5. Flip the ball over in the bowl. Cover with a lid or plate to prevent drying.

Bulk Fermentation (2 to 4 hours)

1. Rest the dough at 74–80°F (23–27°C), performing a set of stretch and folds every 15 to 30 minutes until the dough holds its folded shape somewhat and can make a windowpane several inches wide (usually two or three sets of stretch and folds). Your dough should feel strong and elastic.

2. Continue resting the dough undisturbed until it is full of big and small bubbles, has risen to about twice its original volume, passes the poke test, smells more like bread than flour, and jiggles in the bowl like jelly when you shake it (usually 2 to 4 hours total time after mixing in the levain, depending on dough temperature).

Pre-shape (20 minutes)

1. Turn the dough out onto a lightly misted counter.

2. Fold the dough into thirds, like a letter, with a lightly floured bench knife or spatula. Then fold the other two sides in to make a ball and flip the ball. Rotate it to tighten the skin.

3. Rest 10 to 15 minutes until the ball has relaxed.

Shape the Boule (a couple of minutes)

1. Flour the top of the ball and flip it over. Using lightly floured hands, pull up one edge of the dough and press it into the far side for a couple of seconds to adhere. Continue around the ball once or twice, letting the dough tell you when to stop. When it resists pulling or begins to tear as you are pulling, it is time to stop.

2. Seal the seam by pinching it together and holding until it sticks, then flip the ball over and let it rest a minute to seal.

3. Tighten the skin by dragging the loaf toward you with the bench knife, then rotate and repeat several times. Smooth flour onto the top and sides.

4. Place the loaf seam side up in a well-floured proofing basket. Sprinkle it with flour, especially around the edges. Lay parchment paper on top, then cover it with a lid or place it in a plastic bag so the dough doesn't dry out during proofing.

Proof the Loaf (30 minutes to 24 hours)

Retard by storing in the refrigerator for 4 to 24 hours. Or rest at 74–80°F (23–27°C) until the loaf is risen and jiggly again like a water balloon and passes the poke test (30 to 90 minutes).

DAY 2 OR 3

Bake the Loaf (1½ to 2 hours)

1. Preheat the oven to 475°F (246°C) for 30 to 60 minutes before you want to bake and prepare your baking and steam setup.

2. When the oven is ready, take the dough out from the refrigerator. It should be a little more risen, but don't expect it to be double in size. Holding the parchment paper on top, invert it gently onto your peel or counter.

3. Brush away the excess flour and score the loaf with an X, holding the blade at a 30-degree angle from the dough to cut under the skin rather than deeply into the loaf.

4. Mist the loaf generously (optional) and transfer to the oven using your chosen baking and steam setup.

If baking with a cover: Cover and bake 20 minutes for the steam phase, then uncover and continue baking until the interior is 207°F (97°C) and the crust browns to your liking, usually 35 to 50 minutes total baking time.

If baking with a steam pan: Place the loaf in the oven, then pour 1½ cups (355 milliliters) of water into the steam pan. Bake 20 minutes without opening the oven for the steam phase, then continue baking until the interior is 207°F (97°C) and the crust browns to your liking, usually 35 to 50 minutes total baking time.

5. Cure the crust in the cooling oven with the door ajar for 5 to 10 minutes. Cool for 1 or 2 hours on a wire rack.

Crusty Boule

Take the flavor and texture up a notch with a touch of whole grain wheat flour. We steam the loaf extra long to get a thick crust, which complements the tender, springy crumb. The moderately high-hydration level gives us nice open holes in the crumb.

The whole grain changes the dough a little—you may notice it is a tiny bit less strong in the windowpane test—but it is essentially the same as all-white flour to handle, proof, and bake.

MAKES ONE LOAF

TOTAL FLOUR: 450 GRAMS (1 POUND)

PRE-FERMENTED FLOUR: 17%

HYDRATION: 70%

WHOLE GRAIN: 15%

DOUGH: 774 GRAMS (1.7 POUNDS)

DAY 1

Build the Levain (10 minutes)

1. In a small bowl or pint jar, mix together:

30 grams starter that floats (2 tablespoons)

60 grams water (¼ cup)

60 grams all-purpose flour (½ cup)

2. Cover and let it rest at 60–70°F (16–21°C) for 8 to 12 hours until it is risen and bubbly.

DAY 2

Test the Levain (a couple of minutes)

It should be risen two to three times the original height and bubbly, but not already deflated (a little loss of height is okay). If you drop a spoonful into water, it should float.

Mix the Dough (10 minutes)

1. Mix in a large heavy bowl:

68 grams whole wheat flour (½ cup plus 1 tablespoon)

307 grams bread flour (2½ cups plus 1 tablespoon)

2. Whisk together in the levain bowl or jar until uniform:

240 grams water (1 cup) at 90°F (32°C)

150 grams levain (all)

continued

3. Combine completely with a silicone spatula until all the flour is wet and it forms a ball. This usually takes 2 to 3 minutes. Toward the end, you can use a damp hand to finish mixing, if needed. Do not add additional flour or water.

Autolyse (30 minutes)

1. Cover the bowl with a lid or plate and let the dough rest for 30 minutes at 74–80°F (23–27°C), or until it is relaxed and soft and gives a nice windowpane.

2. Add:

9 grams salt (1½ teaspoons)

3. Sprinkle the salt evenly on the wet surface of the dough and poke and pinch it deep into the dough with your wet fingers.

4. In the bowl, thoroughly mix the salt into the dough with the silicone spatula or a wet hand.

5. Flip the ball over in the bowl. Cover with a lid or plate.

Bulk Fermentation (2 to 4 hours)

1. Rest the dough at 74–80°F (23–27°C), performing stretch and folds every 15 to 30 minutes until the dough holds its folded shape somewhat and can make a windowpane several inches wide (usually two or three sets of stretch and folds). Stop stretch and folds when your dough feels strong and elastic and stays somewhat in its ball shape for the 15 to 30 minute rest after the last stretch and fold.

2. Continue resting undisturbed until the dough is full of big and small bubbles, has risen about twice its original volume, passes the poke test, smells more like bread than flour, and jiggles in the bowl like jelly when you shake it (usually 2 to 4 hours total time after mixing in the levain, depending on dough temperature).

Pre-shape (20 minutes)

1. Turn the dough out onto a lightly misted counter.

2. Fold the dough into thirds, like a letter, with a lightly floured bench knife or spatula. Then fold the other two sides in to make a ball and flip the ball. Rotate it to tighten the skin.

3. Rest 10 to 15 minutes until the ball has relaxed.

Shape the Boule (a couple of minutes)

1. Flour the top of the ball and flip it over. Using lightly floured hands, pull up one edge of the dough and press it into the far side for a couple of seconds to adhere. Continue around the ball once or twice, letting the dough tell you when to stop. When it resists pulling or begins to tear as you are pulling, it is time to stop.

2. Seal the seam by pinching it together and holding it until it sticks, then flip the ball over and let it rest a minute to seal.

3. Tighten the skin by dragging the loaf toward you with the bench knife several times and smooth flour onto the top and sides.

4. Place the loaf seam side up in a well-floured proofing basket. Sprinkle it with flour, especially around the edges. Lay parchment paper on top, then cover it with a lid or place it in a plastic bag so the dough doesn't dry out during proofing.

Proof the Loaf (30 minutes to 18 hours)

Retard the loaf by storing in the refrigerator for 4 to 18 hours. Or let it rest at 74–80°F (23–27°C) until the loaf is risen and jiggly again like a water balloon and passes the poke test (30 to 90 minutes).

DAY 2 OR 3

Bake the Loaf (1½ to 2 hours)

1. Preheat the oven to 475°F (246°C) for 30 to 60 minutes before you want to bake and prepare your baking and steam setup.

2. When the oven is ready, take the dough out from the refrigerator. It should be a little more risen, but don't expect it to be double in size. Holding the parchment paper on top, invert it gently onto your peel or counter.

3. Brush away the excess flour and score the loaf with an X, holding the blade at a 30-degree angle from the dough to cut under the skin rather than deeply into the loaf.

4. Mist the loaf generously (optional) and transfer to the oven using your chosen baking and steam setup.

If baking with a cover: Cover and bake 25 minutes for the steam phase, then uncover and continue baking until the interior is 207°F (97°C) and the crust browns to your liking, usually 35 to 50 minutes total baking time.

If baking with a steam pan: Place the loaf in the oven, then pour 2 cups (474 milliliters) of water into the steam pan. Bake 25 minutes without opening the oven for the steam phase, then continue baking until the interior is 207°F (97°C) and the crust browns to your liking, usually 35 to 50 minutes total baking time.

5. Cure the crust in the cooling oven with the door ajar for 5 to 10 minutes. Cool for 1 or 2 hours on a wire rack.

Rustic Boule

This bread strikes a wonderful balance between the flavor and healthfulness of whole wheat and the delight of an open crumb and springy interior. The big, fresh levain and short bulk fermentation make this bread only mildly sour for a whole grain sourdough. An overnight retard, if you have time, affords sufficient fermentation and a lovely oven spring.

A whole grain dough behaves very differently than white. It needs gentle handling due to the bran in the flour, and the fermentation temperature is best kept below 80°F (27°C). When it starts to look puffy, check frequently for it to be a little bit jiggly, then promptly end the bulk fermentation. We lower the baking temperature to prevent over-browning. The barley malt flour is not required, but it helps release the maltose sugars that yeast loves and will result in a fluffier crumb.

MAKES ONE LOAF

TOTAL FLOUR: 450 GRAMS (1 POUND)
PRE-FERMENTED FLOUR: 17%
HYDRATION: 80%
WHOLE GRAIN: 72%
DOUGH: 819 GRAMS (1.8 POUNDS)

DAY 1

Build the Levain (10 minutes)

1. In a small bowl or pint jar, mix together:

30 grams starter that floats (2 tablespoons)

60 grams water (¼ cup)

60 grams whole wheat flour (½ cup)

2. Cover and let it rest at 60–70°F (16–21°C) for 8 to 12 hours until it is risen and bubbly.

DAY 2

Test the Levain (a couple of minutes)

It should be risen two to three times the original height and bubbly, but not already deflated (a little loss of height is okay). If you drop a spoonful into water, it should float.

Mix the Dough (10 minutes)

1. Mix in a large heavy bowl:

256 grams whole wheat flour (2 cups plus 2 tablespoons)

110 grams bread flour (scant 1 cup)

9 grams diastatic barley malt flour (1 tablespoon; optional)

continued

2. Whisk together in the levain bowl or jar until uniform:

285 grams (1¼ cups plus 1 tablespoon) water at 90°F (32°C)

150 grams ready levain (all)

3. Mix with a silicone spatula or your hand until it is combined completely and all the flour is wet.

Autolyse (1 hour)

1. Cover the bowl with a lid or plate and rest the dough for an hour at 74–80°F (23–27°C) until the bran is wet and softened and the dough is relaxed and soft and can make a small windowpane.
2. Add:

9 grams salt (1½ teaspoons)

3. Sprinkle the salt evenly on the wet surface of the dough and poke and pinch it deep into the dough with your wet fingers.
4. Thoroughly mix the salt into the dough in the bowl with the silicone spatula or a wet hand.
5. Flip the ball over in the bowl. Cover with a lid or plate.

Bulk Fermentation (2 to 4 hours)

1. Rest the dough at 74–80°F (23–27°C), performing a set of very gentle stretch and folds every 15 to 30 minutes until the dough holds its folded shape somewhat and can make a 1- to 2-inch windowpane (usually only one or two sets). It won't feel as strong as a white flour dough. Extra handling of the dough should be avoided at this point forward because the bran damages the gluten network during handling.
2. Continue resting the dough undisturbed until it is risen and seems light, with bubbles throughout, the smell of flour is gone and it smells more like bread, and the dough just begins to jiggle in its bowl when you shake it (usually 2 to 4 hours total time after mixing in the levain, depending on the dough temperature).

Pre-shape (20 minutes)

1. Turn the dough out onto a lightly misted counter.
2. Fold the dough into thirds, like a letter, with a lightly floured bench knife or spatula. Then fold the other two sides in to make a ball. Flip the ball and rotate it to tighten the skin.
3. Rest 15 minutes until the ball has relaxed.

Shape the Boule (a couple of minutes)

1. Flour the top of the ball and flip it over. Using lightly floured hands, gently pull up one edge of the dough and press it into the far side for a couple of seconds to adhere. Continue around the ball while carefully watching for any tearing, letting the dough tell you when to stop. When it resists pulling or begins to tear as you are pulling, it is time to stop. With a whole grain dough, the ball will be looser than what is possible with an all-white dough because of the bran cutting into the gluten network when the dough is handled.
2. Seal the seam by pinching it together and holding until it sticks, then flip the ball over and let it rest a minute to seal.
3. Tighten the skin by slowly dragging the loaf toward you with the bench knife, being very careful not to tear the skin, and smooth flour on the top and sides.
4. Place the loaf seam side up in a well-floured proofing basket. Sprinkle it with flour, especially around the edges. Lay parchment paper on top, then cover it with a lid, or place it in a plastic bag so the dough doesn't dry out during proofing.

Proof the Loaf (30 minutes to 12 hours)

Retard the loaf by storing in the refrigerator for 4 to 12 hours. Or let it rest at 74–80°F (23–27°C) until the loaf is risen and jiggly again like a water balloon and passes the poke test (30 to 90 minutes).

DAY 2 OR 3

Bake the Loaf (1½ to 2 hours)

1. Preheat the oven to 450°F (232°C) for 30 to 60 minutes before you want to bake and prepare your baking and steam setup.
2. When the oven is ready, take the dough out from the refrigerator. It should be a little more risen, but don't expect it to be double in size. Holding the parchment paper on top, invert it gently onto your peel or counter.
3. Brush away the excess flour and score the loaf with one cut, holding the blade at a 30-degree angle from the dough to cut under the skin rather than deeply into the loaf.
4. Mist the loaf generously and transfer to the oven using your chosen baking and steam setup.

If baking with a cover: Cover and bake 25 minutes for the steam phase, then uncover and continue baking until the interior is 207°F (97°C) and the crust browns to your liking, usually 35 to 50 minutes total baking time.

If baking with a steam pan: Place the loaf in the oven, then pour 2 cups (474 milliliters) of water into the steam pan. Bake 25 minutes without opening the oven for the steam phase, then continue baking until the interior is 207°F (97°C) and the crust browns to your liking, usually 35 to 50 minutes total baking time.

5. Cure the crust in the cooling oven with the door ajar for 5 to 10 minutes. Cool for 1 or 2 hours on a wire rack.

Octo Rolls

Rolls are a fun way to enjoy some sourdough. They are great for sandwiches, soups, and stews, sharing with a friend, or freezing. Start with one of the three Essential Recipes and follow the recipe through the end of the bulk fermentation step. Give yourself lots of space on the counter so you have room to shape and rest all these little puppies.

In this recipe, we reduce the steaming time and oven temperature for a thinner, softer crust than on the boule. Simple Boule dough was used for the photographs.

MAKES 8 ROLLS

Pre-shape (15 minutes)

1. Line an 18-by-13-inch (46-by-33-centimeter) baking sheet with parchment paper (or place the parchment paper on a peel if you want to bake on a stone).

2. Turn dough out onto a lightly misted counter.

3. Divide the dough in half using a bench knife. Divide each half into four even pieces so you have a total of eight pieces. Use a scale if you really want them to be exactly the same size.

4. Pre-shape each piece into a ball by folding the dough into thirds, like a letter, with a lightly floured bench knife or spatula. Then fold the other two sides in to make a ball. Flip the ball and rotate it to tighten the skin.

5. Rest 10 minutes until the ball has relaxed.

Shape the Rolls (10 minutes)

1. Flour the top of each ball and flip it over. Using lightly floured hands, gently pull up one edge of the dough and press it into the far side for a couple of seconds to adhere. Continue around the ball while carefully watching for any tearing, letting the dough tell you when to stop. When it resists pulling or begins to tear as you are pulling, it is time to stop. With a whole grain dough, the ball will be looser than what is possible with an all-white dough because of the bran cutting into the gluten network when the dough is handled.

2. Seal the seam by pinching it together and holding until it sticks, then flip the ball over and let it rest a minute to seal.

3. Tighten the skin by slowly dragging the roll toward you with the bench knife, being very careful not to tear the skin.

4. Place the rolls seam side down on the parchment paper with room to expand between them.

continued

5. Cover with parchment paper and a damp towel or puffed-up plastic bag to keep them from drying and let it rest at 74–80°F (23–27°C) until well risen and puffy (30 to 90 minutes) or retard by storing them in the refrigerator for 3 to 12 hours.

Bake the Rolls (1½ hours)

1. Preheat the oven to 425°F (218°C) for 30 to 60 minutes before you want to bake and prepare your baking and steam setup.

2. When the oven is ready, score the rolls with an X and mist them heavily with water (optional).

3. Place the pan in the oven (or slide them, parchment paper and all, onto a hot baking stone).

If baking with a cover: Cover and bake 10 minutes for the steam phase, then uncover and continue baking until the interior is 207°F (97°C) and the crust browns to your liking, usually 20 to 35 minutes total baking time.

If baking with a steam pan: Pour 1 cup (237 milliliters) of water into the steam pan. Bake 10 minutes without opening the oven for the steam phase, then continue baking until the interior is 207°F (97°C) and the crust browns to your liking, usually 20 to 35 minutes total baking time.

4. Cool on a rack for at least 15 minutes before serving.

CHAPTER 7

Following Your Muse: Personalize the Essential Loaf

Once you've become comfortable with your chosen Essential Recipe, it's time to branch out! You are ready to try the recipes in the following chapters. However, you may also want to customize the fantastic loaves you can now bake using the Essential Recipes. Here are some ideas.

Easy: Seeds on the Crust

Poppy, sesame, flax, sunflower, caraway, and even rolled oats can be pressed into your crust if you wish. Prepare a piece of waxed paper with a sprinkling of seeds. After you shape your loaf, do not sprinkle it with flour. Rub or spray the entire top and sides completely with oil (to prevent sticking to the paper) and lay the loaf, top side down, onto the seeds. Lift the waxed paper and dough into the proofing basket. Sprinkle a few more seeds around the edges where the dough likes to stick as it rises. After proofing and turning out the loaf, peel off the waxed paper before baking. Keep watch on the seeds toward the end and tent the loaf with foil if they are browning too soon.

Easy: Spice It Up

Fragrances from your spice drawer or garden like cinnamon, rosemary, orange zest, anise, dill, basil, and thyme are enchanting in a loaf. Some herbs and spices will interfere with the growth of yeast and should not be added to the dough when mixing. For example, large amounts of cinnamon, garlic, and ginger slow the growth of yeast. If

you're not sure if the spice you want to use will affect the dough, sprinkle it on the dough after turning out the bulk fermentation, and it will become folded into the dough during the usual shaping process.

Intermediate: Express Yourself with Your Lame

Scoring the bread helps it rise in the oven, but scoring can also be used to make art on the crust. You've already learned to score through the dough's skin without cutting into the layers underneath. Now experiment with making additional, very shallow cuts that do not go all the way through the skin to create a design on the crust. You can pair these shallow cuts with your main score in myriad ways, creating a new pattern on every loaf or designing your own signature scoring pattern. Rubbing the loaf evenly with flour helps the patterns stand out (don't mist the loaf in this case). Chilling the dough before scoring makes it easier to be exact when cutting.

Intermediate: Studded Bread

Loaves studded with olives, dried fruits, nuts, chocolate chips, cheese, or roasted garlic cloves are delicious. These ingredients can be added to the dough with the salt. They will usually cause a denser loaf than the original recipe. Start with 1 cup per loaf and see how you like it. Dried fruits steal some water from the dough, so if you don't want a closer crumb, you can either soak and drain the fruit before using or add them after the bulk fermentation. During the usual shaping process they will be distributed in the dough. Wet foods, like olives, may add hydration to the dough, so be prepared to add extra flour (just a few teaspoons) if the dough seems wetter than normal after mixing. Shaping the loaf will be more challenging, but it is certainly worth the trouble.

Advanced: Grain Adventures

The flour options are stunning: spelt, rye, khorasan, durum, millet, buckwheat, emmer, quinoa, amaranth, and sprouted grain flours. There are so many exciting flours available, each with its own flavors, nutrition profile, and behavior in the dough. There are ancient grains, wheat relatives, and non-wheat grains from around the world. Some of these have gluten proteins and will help hold up the dough, others lack gluten and will make a denser loaf. In any of the Essential Recipes you can swap out one-tenth of the flour for a different flour and get a successful loaf that will still rise nicely. You might even create your own personal favorite. Keep notes on the dough using a worksheet so you can replicate a great result or tweak one that could use improvement.

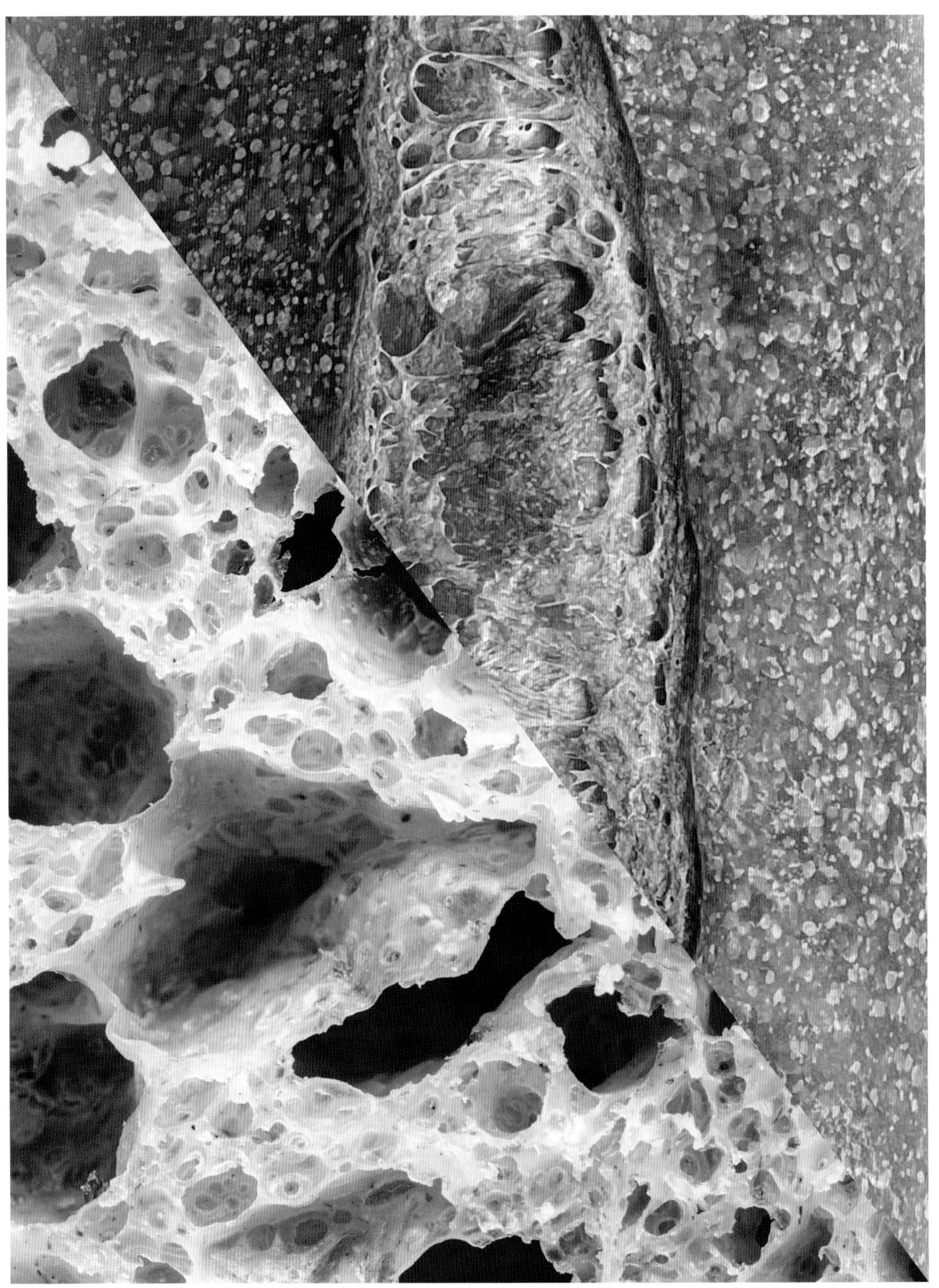

CHAPTER 8

Tour de Force: Recipes Made with a Levain

These are the big shoulders—the recipes that get the work done, the main go-to recipes. You can control many different aspects of each recipe to your preference. The three Essential Recipes from Chapter 6 used levains, so these recipes will have a familiar rhythm. The nice thing about using a levain is the dependability of the fermentation. Since you know you have fresh peak levain yeast in your dough, you can predict how the process will go based on temperature. Also, breads made with a levain can undergo a pretty fast bulk fermentation and can be baked up the same day, leading to a mild lactic acid content in the finished bread. This is great if you don't want a sour-flavored bread. If you do want sour flavor, check out the Golden Gate Gem recipe (page 147), where we use several techniques to amplify tartness.

This chapter takes you through a diverse set of recipes demonstrating the possibilities for creativity in your bread baking. Many feature some whole or sprouted grains and other ingredients. You can bake these recipes as written or modify them with the ideas from Chapter 7 or in the recipe description. None of these recipes will be challenging if you are already comfortable with baking the Essential Recipes; however, they are arranged with the easier recipes at the beginning of the chapter and the more tricky recipes at the end.

Maximizing Flavor

Although a sourdough starter can make bread taste sour, it doesn't have to, and some prefer the lovely complexity of flavors imparted by wild yeast without a strong acid backdrop. An overnight retard will yield a more flavorful bread, and yet the bread still won't be very sour in these levain recipes. Extra fermentation time in the refrigerator adds to the flavor of the bread because the sourdough organisms slow down and make interesting flavor compounds that accumulate. Think of aging a wine or cheese to develop flavor—aging the dough, likewise, helps add to the complexity of the flavor. Normally this includes tangy acetic acid. However, using a fresh levain and keeping the bulk fermentation at yeast's ideal temperature greatly curbs the population of the bacteria that can produce acetic acid in the dough, especially if you use a recently refreshed starter to build your levain. The other flavor compounds will accumulate, as the wild yeast continue to perform at cooler temperatures.

Balancing Bulk Fermentation and Loaf Proofing for Optimal Rising

When the dough gets to the oven, it should have enough fermentation overall to become bread rather than a dense brick, but not so much that it becomes a flat flying saucer. The total fermentation time is divided between the bulk fermentation and the loaf proofing steps. To keep the dough in the bread window, keep the total amount of fermentation in mind. If you take your bulk fermentation very far, so that the dough has tripled or quadrupled in size and is very jiggly, then use a very short loaf proof, baking your loaf as soon as the oven is hot (thus avoiding a flying saucer). If you end up forming the loaf while the dough is just barely doubled in volume and not jiggly, then give the loaf the full amount of time to proof or an overnight retard to complete the fermentation (and avoid a brick). Remember, too, that temperature has a powerful influence on the speed of fermentation. A fermentation at 80°F (27°C) will need less time than one at a lower temperature.

Fitting the Recipe to Your Schedule

The recipes are scheduled over 2 or 3 days, but the first day all you are doing is building the levain, which takes 10 minutes. If you want to speed things up, you can opt to build a fast levain on the same day you mix the dough. You can retard the dough at either the bulk fermentation or the loaf proof. For retarding the bulk fermentation, it's best to get all your stretch and folds in and to let the rising start before placing the dough in the refrigerator. For retarding the loaf, refrigerate it as soon as it's shaped and put it right into the oven from the refrigerator.

RHYTHM 1. AT HOME IN THE AFTERNOON, BREAD FOR LATE DINNER OR TOMORROW'S BREAKFAST. Day 1: In the morning, build a fast 1:1:1 levain instead of the 1:2:2 levain in the recipe. To do this, simply mix the levain using 50 grams each of starter, water, and flour, and rest at 74–80°F (23–27°C). When it's ready in a few hours, mix up your dough and tend to it periodically over several hours. Shape the loaf and bake it in the evening to have for a late dinner or the next morning, or retard the loaf overnight and bake it the following morning (Day 2).

RHYTHM 2. OUT ALL DAY, BAKE ANYTIME TOMORROW. Day 1: Build the levain the first thing in the morning. When it's ready in 8 hours, mix up your dough and tend to it periodically over several hours in the evening. You can place the dough in the refrigerator and finish the recipe the next day. Or you can shape the loaf and retard it overnight, then bake it the following day (Day 2).

RHYTHM 3. AT HOME IN THE MORNING, BREAD FOR TONIGHT'S DINNER OR TOMORROW'S BREAKFAST. Day 1: Build the levain just before bed. When it's ready the first thing in the morning on Day 2, mix up your dough and tend to it periodically over several hours. Shape the loaf and bake it that afternoon to have for dinner or for breakfast the next morning. Or you could retard the loaf for several hours in the refrigerator and bake later that day, or you could retard the loaf overnight and bake the following morning on Day 3.

Levain: 10 minutes to mix, proof for 2 hours (fast levain) up to 12 hours (regular levain)

- Mixing the dough: 10 minutes
- Autolyse: 30 minutes
- Bulk fermentation: 2 to 4 hours
- Preshape: 20 minutes
- Proof: 20 minutes to 24 hours
- Bake: 1 to 2 hours

Faithful Batard

A nice choice for a standard go-to bread, this versatile loaf tends to please everyone and is easy to bake. With a touch of whole wheat and rye, the Faithful Batard shines with flavor. It rises well and features a medium open crumb and a crunchy crust. Pair it with cheeses or soups, take it on a picnic, or make it into sandwiches.

MAKES ONE LOAF

TOTAL FLOUR: 450 GRAMS (1 POUND)

PRE-FERMENTED FLOUR: 17%

HYDRATION: 75%

WHOLE GRAIN: 25%

DOUGH: 797 GRAMS (1.7 POUNDS)

DAY 1

Build the Levain (10 minutes)

1. In a small bowl or pint jar, mix together:

30 grams starter that floats (2 tablespoons)

60 grams water (¼ cup)

60 grams all-purpose flour (½ cup)

2. Cover and let it rest at 60–70°F (16–21°C) for 8 to 12 hours until it is risen and bubbly and can float in water.

DAY 2

Mix the Dough (10 minutes)

1. Mix the flour in a large heavy bowl and set aside:

75 grams whole wheat flour (⅝ cup)

38 grams whole rye flour (¼ cup plus 1 tablespoon)

262 grams bread flour (2¼ cups)

2. Whisk together in the levain bowl or jar until uniform:

263 grams water (1⅛ cups) at 90°F (32°C)

150 grams levain (all)

3. Mix the wet ingredients into the flour, combining completely.

Autolyse (30 to 60 minutes)

1. Rest the dough for 30 to 60 minutes until it has relaxed in the bowl.

2. Add:

9 grams salt (1½ teaspoons)

continued

3. Sprinkle the salt over the wet dough and mix it in by poking and cutting it into the dough with a wet spatula, then folding the edges of the dough over the center with a spatula until the dough resists folding across itself, about 20 folds. Flip the ball of dough in the bowl and cover.

Bulk Fermentation (3 to 5 hours)

1. Rest the dough at 74–80°F (23–27°C), with two to four sets of stretch and folds 15 to 30 minutes apart until the dough holds its ball shape somewhat and makes a nice windowpane.

2. Let it rise until the dough is airy and puffy, double or triple or more in size, very jiggly, and smells like bread (3 to 5 hours total time after mixing in the levain).

Shape the Loaf (20 minutes)

1. Handle the dough minimally and gently during shaping to achieve a light, lofty crumb. Turn the dough out onto a lightly misted counter and pre-shape: fold the dough into thirds, like a letter, fold in the other two sides, and flip the ball. Rotate it to tighten the skin. Let it rest 10 to 15 minutes until it relaxes.

2. Shape into a batard (or boule) and gently tighten the skin, being careful not to tear the skin. Flour the top generously.

3. Place the loaf seam side up in a well-floured proofing basket. Sprinkle it with flour, especially around the edges. Lay parchment paper on top.

Proof the Loaf (1 to 18 hours)

Cover and retard the loaf in the refrigerator for 4 to 18 hours. Or let it rest at 74–80°F (23–27°C) until risen and jiggly like a water balloon (1 to 3 hours).

DAY 2 OR 3

Bake the Loaf

1. Preheat the oven to 475°F (246°C) 30 to 60 minutes before baking, and prepare your baking and steam setup.

2. Turn out the loaf and score it.

3. Mist the loaf generously (optional) and transfer to the oven using your chosen baking and steam setup.

If baking with a cover: Cover and bake 25 minutes for the steam phase, then uncover and continue baking until the interior is 207°F (97°C) and the crust browns to your liking, usually 35 to 50 minutes total baking time.

If baking with a steam pan: Place the loaf in the oven, then pour 1½ cups (355 milliliters) of water into the steam pan. Bake 25 minutes without opening the oven for the steam phase, then continue baking until the interior is 207°F (97°C) and the crust browns to your liking, usually 35 to 50 minutes total baking time.

4. Cure in the cooling oven with the door ajar for 5 to 10 minutes. Cool 1 to 2 hours on a rack.

Heavenly High Hydration

This recipe is a great way to build up your high-hydration dough skills while enjoying a fantastic loaf of bread. At 75 percent hydration, this all-white-flour dough will be pretty slack and sticky, but the bread is definitely worth the effort. A blistered redwood-toned crust and an open, springy, delectable crumb are your rewards for the trouble.

MAKES ONE LOAF

TOTAL FLOUR: 450 GRAMS (1 POUND)

PRE-FERMENTED FLOUR: 17%

HYDRATION: 75%

WHOLE GRAIN: 0%

DOUGH: 797 GRAMS (1.8 POUNDS)

DAY 1

Build the Levain (10 minutes)

1. In a small bowl or pint jar, mix together:

30 grams starter that floats (2 tablespoons)

60 grams water (¼ cup)

60 grams all-purpose flour (½ cup)

2. Cover and let it rest at 60–70°F (16–21°C) for 8 to 12 hours until it is risen and bubbly and can float in water.

DAY 2

Mix the Dough (10 minutes)

1. Measure the flour into a large heavy bowl and set aside:

375 grams bread flour (3⅛ cups)

2. Whisk together in the levain bowl or jar until uniform:

263 grams water (1⅛ cups) at 90°F (32°C)

150 grams levain (all)

3. Mix the wet ingredients into the flour, combining completely. The dough will seem very wet and sticky, not forming a ball at all.

Autolyse (30 to 60 minutes)

1. Rest the dough for 30 to 60 minutes until the dough has relaxed in the bowl.

2. Add:

9 grams salt (1½ teaspoons)

continued

3. Sprinkle the salt over the wet dough and mix it in by poking and cutting it into the dough with a wet spatula, then folding the edges of the dough over the center with a spatula until the dough resists folding across itself, about 20 folds. Flip the ball of dough in the bowl and cover.

Bulk Fermentation (3 to 5 hours)

1. Rest the dough at 74–80°F (23–27°C), with three or four sets of stretch and folds 15 to 30 minutes apart until the dough holds its ball shape somewhat and makes a nice windowpane. For this wet dough, I prefer coil folds.

2. Let it rise until the dough is airy and puffy, double or triple in size, very jiggly, and smells like bread (3 to 5 hours total time after mixing in the levain).

Shape the Loaf (20 minutes)

1. Handle the dough minimally and gently during shaping to achieve a light, lofty crumb. Turn the dough out onto a lightly misted counter and pre-shape: fold the dough into thirds, like a letter, fold in the other two sides, and flip the ball. Rotate it to tighten the skin. Let it rest 10 to 15 minutes until it relaxes.

2. Shape into a batard (or boule) and gently tighten the skin, being careful not to tear the skin. Flour the top generously.

3. Place the loaf seam side up in a well-floured proofing basket. Sprinkle it with flour, especially around the edges. Lay parchment paper on top.

Proof the Loaf (1 to 24 hours)

Cover and retard the loaf in the refrigerator for 4 to 24 hours. Or let it rest at 74–80°F (23–27°C) until risen and jiggly like a water balloon (1 to 3 hours).

DAY 2 OR 3

Bake the Loaf

1. Preheat the oven to 500°F (260°C) 30 to 60 minutes before baking, and prepare your baking and steam setup.

2. Turn out the loaf and score it.

3. Mist the loaf generously (optional) and transfer to the oven using your chosen baking and steam setup.

If baking with a cover: Cover and bake 25 minutes for the steam phase, then uncover and continue baking until the interior is 207°F (97°C) and the crust browns to your liking, usually 35 to 50 minutes total baking time.

If baking with a steam pan: Place the loaf in the oven, then pour 1½ cups (355 milliliters) of water into the steam pan. Bake 25 minutes without opening the oven for the steam phase, then continue baking until the interior is 207°F (97°C) and the crust browns to your liking, usually 35 to 50 minutes total baking time.

4. Cure in the cooling oven with the door ajar for 5 to 10 minutes. Cool 1 to 2 hours on a rack.

Herb and Cheese Bread

This addictive, flavorful bread makes an amazing foundation for a ham sandwich or pan con tomate. The delightful nuggets of melted cheese throughout the bread are irresistible—and the smell while it's baking will pique the appetite of anyone with a pulse. A dash of whole wheat gives extra flavor to the crumb, to help it stand up to the powerful cheese and herbs. Here we use oregano, thyme, and Asiago, but you can make this bread your own. Substitute any fresh herbs such as dill, chives, basil, parsley, sage, rosemary, or tarragon. Swap the Asiago for any firm, aged, sharp cheese like cheddar, Gruyère, Gouda, or manchego.

MAKES ONE LOAF

TOTAL FLOUR: 450 GRAMS (1 POUND)

PRE-FERMENTED FLOUR: 17%

HYDRATION: 80%

WHOLE GRAIN: 20%

DOUGH: 1001 GRAMS (2.2 POUNDS)

DAY 1

Build the Levain (10 minutes)

1. In a small bowl or pint jar, mix together:

> **30 grams starter that floats (2 tablespoons)**
>
> **60 grams water (¼ cup)**
>
> **60 grams all-purpose flour (½ cup)**

2. Cover and let it rest at 60–70°F (16–21°C) for 8 to 12 hours until it is risen and bubbly and can float in water.

DAY 2

Mix Dough (10 minutes)

1. Mix the flours in a large heavy bowl and set aside:

> **285 grams bread flour (2⅜ cups)**
>
> **90 grams whole wheat flour (¾ cup)**

2. Whisk together in the levain bowl or jar until uniform:

> **150 grams levain (all)**
>
> **285 grams water (1 cup plus 3 tablespoons) at 90°F (32°C)**

3. Mix the wet ingredients into the flour, combining completely.

continued

Autolyse (30 to 60 minutes)

1. Rest the dough for 30 to 60 minutes until the dough has relaxed in the bowl.
2. Add:

9 grams salt (1½ teaspoons)

3. Sprinkle the salt over the wet dough and mix it in by poking and cutting it into the dough with a wet spatula, then folding the edges of the dough over the center with a spatula until the dough resists folding across itself, about 20 folds. Flip the ball of dough in the bowl and cover.

Bulk Fermentation (3 to 5 hours)

1. Rest the dough at 74–80°F (23–27°C), with one or two sets of stretch and folds 15 to 30 minutes apart until the dough holds its ball shape somewhat and gives a small windowpane.
2. Sprinkle evenly onto the dough:

6 grams fresh thyme leaves (2 tablespoons) from 8 to 10 sprigs

6 grams fresh oregano leaves (2 tablespoons) from 4 to 6 sprigs

170 grams cubed Asiago cheese (6 ounces), warmed to room temperature

3. Perform a final set of stretch and folds to fold the herbs and cheese into the dough. Then let it rise until the dough is airy and puffy, double or triple in size, and becoming jiggly (3 to 5 hours total time after mixing in the levain).

Shape the Loaf (20 minutes)

1. Handle the dough minimally and gently during shaping to achieve a light, lofty crumb. Turn the dough out onto a lightly misted counter and pre-shape: fold the dough into thirds, like a letter, fold in the other two sides, and flip the ball. Rotate it to tighten the skin. Let it rest 10 to 15 minutes until it relaxes.

2. Shape into a boule or batard and gently tighten the skin, watching closely for tearing. Try to poke any escaping cheese cubes back into the loaf. Flour the top generously.

3. Place the loaf seam side up in a well-floured proofing basket. Sprinkle it with flour, especially around the edges. Lay parchment paper on top.

Proof the Loaf (1 to 12 hours)

Cover and retard the loaf in the refrigerator for 4 to 12 hours. Or let it rest at 74–80°F (23–27°C) until risen and jiggly like a water balloon (1 to 3 hours).

Bake the Loaf

1. Preheat the oven to 475°F (246°C) 30 to 60 minutes before baking, and prepare your baking and steam setup.

2. Turn out the loaf and score it.

3. Mist the loaf generously (optional) and transfer to the oven using your chosen baking and steam setup.

If baking with a cover: Cover and bake 25 minutes for the steam phase, then uncover and continue baking until the interior is 207°F (97°C) and the crust browns to your liking, usually 35 to 50 minutes total baking time.

If baking with a steam pan: Place the loaf in the oven, then pour 1½ cups (355 milliliters) of water into the steam pan. Bake 25 minutes without opening the oven for the steam phase, then continue baking until the interior is 207°F (97°C) and the crust browns to your liking, usually 35 to 50 minutes total baking time.

4. Cure in the cooling oven with the door ajar for 5 to 10 minutes. Cool 1 to 2 hours on a rack.

Winter Comfort Bread

Roasted garlic, black pepper, and sage join winter squash in a flavorful and colorful combination. Bake this when the vibrant winter squashes start filling the produce aisle and the air gets a nip that makes you want to put on a sweater. It pairs well with winter comfort foods like mushroom ragout or cheese fondue, and it dazzles when used to make the Thanksgiving stuffing or turkey sandwiches.

Roasting the squash brings out ambrosial flavors as the natural sugars caramelize. Choose any kind of starchy winter squash. Bake up a baby butternut, acorn, pumpkin, or delicata squash just for the occasion. Or you can use the eyes, noses, and mouths cutouts of your jack-o'-lantern pumpkins, or squash leftovers from a meal.

MAKES ONE LOAF

TOTAL FLOUR: 450 GRAMS (1 POUND)

PRE-FERMENTED FLOUR: 17%

HYDRATION: 65%

WHOLE GRAIN: 13%

DOUGH: 1002 GRAMS (2.2 POUNDS)

DAY 1

Build the Levain (10 minutes)

1. In a small bowl or pint jar, mix together:

   30 grams starter that floats (2 tablespoons)

   60 grams water (¼ cup)

   60 grams whole wheat flour (½ cup)

2. Cover and let it rest at 60–70°F (16–21°C) for 8 to 12 hours until it is risen and bubbly and can float in water.

DAY 1 OR 2

Prepare the Winter Squash and the Garlic (1 to 1½ hours)

1 small winter squash, cut in half lengthwise, seeds and pith removed

olive oil

1 head of garlic, outer skin removed and top ½ inch cut off to expose cloves (1 centimeter)

1. Preheat oven to 400°F (204°C).
2. Place the squash cut side down in a baking dish.
3. Drizzle olive oil onto the cut surface of the garlic head and wrap the garlic loosely in foil. Place it in the baking dish next to the squash.
4. Put the pan in the oven and roast, uncovered.

continued

5. Remove the garlic when the cloves feel tender, 30 to 40 minutes. Cool, then release the cloves from the skins. Halve or quarter the cloves. Cover and refrigerate if not using immediately, warming to room temperature before using.

6. Remove the squash when the skin and the cut surface are caramelized and brown and the flesh is very soft, 40 to 60 minutes. Cool, then remove the skin. If the squash is very watery, allow it to drain in a colander, pressing out the liquid (which can be used to replace an equal amount of water when mixing the dough). Mash or puree until smooth without adding any liquid. Cover and refrigerate if not using immediately, warming to room temperature before using.

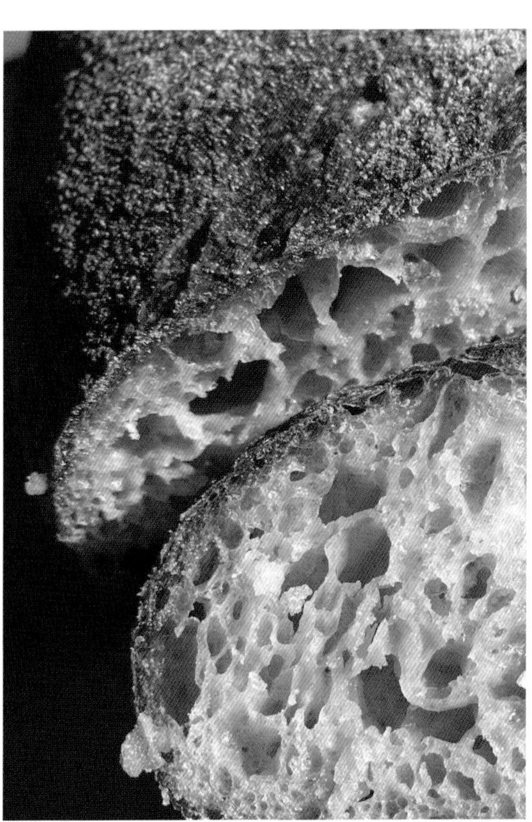

DAY 2

Mix the Dough (10 minutes)

1. Measure the flour into a large heavy bowl and set aside:

375 grams bread flour (3⅛ cups)

2. Whisk together in a medium bowl until uniform:

218 grams water (⅞ cup) at 90°F (32°C)

225 grams of the prepared winter squash (scant 1 cup) at room temperature

150 grams levain (all)

3. Mix the wet ingredients into the flour, combining completely. The dough will feel very wet but should not feel loose. If it is soupy, which may happen if you had a very wet squash, add a tablespoon of flour at a time and mix it in until it feels very soft and sticky but holds together.

Autolyse (30 to 60 minutes)

1. Rest the dough for 30 to 60 minutes until the dough has relaxed in the bowl.

2. Add:

9 grams salt (1½ teaspoons)

1 gram dried sage (2 teaspoons) or 6 grams fresh sage chiffonade (2 tablespoons)

prepared roasted garlic at room temperature

3. Sprinkle the salt, sage, and garlic over the wet dough and mix it in by poking and cutting it into the dough with a wet spatula, then folding the edges of the dough over the center with a spatula until the dough resists folding across itself, about 30 folds. Flip the ball of dough in the bowl and cover.

Bulk Fermentation (3 to 5 hours)

1. Rest the dough at 74–80°F (23–27°C), with two or three sets of stretch and folds 15 to 30 minutes apart until the dough holds its ball shape somewhat and makes a nice windowpane.

2. Let it rise until the dough is airy and puffy, double or triple in size and jiggly (3 to 5 hours total time after mixing in the levain).

Shape the Loaf (20 minutes)

Handle the dough minimally and gently during shaping to achieve a light, lofty crumb. Turn the dough out onto a lightly misted counter and pre-shape: fold the dough into thirds, like a letter, fold in the other two sides, and flip the ball. Rotate it to tighten the skin. Let it rest 10 to 15 minutes until it relaxes.

3. Shape into a batard or boule and gently tighten the skin, being careful not to tear the skin. Flour the top generously.

4. Place the loaf seam side up in a well-floured proofing basket. Sprinkle it with flour, especially around the edges. Lay parchment paper on top.

Proof the Loaf (1 to 18 hours)

Cover and retard the loaf in the refrigerator for 4 to 18 hours. Or let it rest at 74–80°F (23–27°C) until risen and jiggly like a water balloon (1 to 3 hours).

DAY 2 OR 3

Bake the Loaf

1. Preheat the oven to 475°F (246°C) 30 to 60 minutes before baking, and prepare your baking and steam setup.

2. Turn out the loaf, brush off excess flour, and lightly mist it. Sprinkle the top of the loaf with:

freshly milled black pepper to taste

salt to taste—not too much

3. Score the loaf and transfer to the oven using your chosen baking and steam setup.

If baking with a cover: Cover and bake 25 minutes for the steam phase, then uncover and continue baking until the interior is 207°F (97°C) and the crust browns to your liking, usually 35 to 50 minutes total baking time.

If baking with a steam pan: Place the loaf in the oven, then pour 1½ cups (355 milliliters) of water into the steam pan. Bake 25 minutes without opening the oven for the steam phase, then continue baking until the interior is 207°F (97°C) and the crust browns to your liking, usually 35 to 50 minutes total baking time.

4. Cure in the cooling oven with the door ajar for 5 to 10 minutes. Cool 1 to 2 hours on a rack.

Pony Bread

This recipe calls for whole oats, known as oat groats, which we will soak overnight and mix into the dough with the salt. Yes, oat groats are what they feed ponies, but these oats are so, so good, imparting their creamy aroma and springy texture to the bread. This soft, springy bread will stick to your ribs, energizing you for hours.

You can substitute steel cut oats or millet for the oat groats. For a fun breakfast bread, try adding some grated apples, raisins, or cinnamon to the dough after turning it out.

MAKES ONE LOAF

TOTAL FLOUR: 450 GRAMS + 100 GRAMS OATS (1 POUND PLUS 0.2 POUND)

PRE-FERMENTED FLOUR: 17%

HYDRATION: 68%

WHOLE GRAIN FLOUR: 0%

TOTAL WHOLE GRAIN: 25%

DOUGH: ABOUT 1000 GRAMS (2.2 POUNDS)

DAY 1

Soak the Oats (5 minutes)

1. In a small bowl or pint jar, mix together:

100 grams oat groats or steel cut oats (½ cup)

about 350 grams water (1½ cups)

2. Rinse and drain the grains, then cover with the water. Cover and let it rest on the counter overnight.

Build the Levain (10 minutes)

1. In a small bowl or pint jar, mix together:

30 grams starter that floats (2 tablespoons)

60 grams water (¼ cup)

60 grams all-purpose flour (½ cup)

2. Cover and let it rest at 60–70°F (16–21°C) for 8 to 12 hours until it is risen and bubbly and can float in water.

continued

DAY 2

Mix the Dough (10 minutes)

1. Measure the flour into a large heavy bowl and set aside:

375 grams bread flour (3⅛ cups)

2. Whisk together in the levain bowl or jar until uniform:

150 grams levain (all)

231 grams water (scant 1 cup) at 90°F (32°C)

3. Mix the wet ingredients into the flour, combining completely.

Autolyse (30–60 minutes)

1. Rest the dough for 30 to 60 minutes at 74–80°F (23–27°C) until the dough is very relaxed in the bowl to help with incorporating the grains.

2. Rinse the grains under running water and leave to drain completely, blotting the strainer with a towel just before use.

3. Sprinkle onto the surface of the dough:

9 grams salt (1½ teaspoons)

all the soaked, drained oats

4. Poke and pinch the salt and grains into the dough. Thoroughly mix the dough in the bowl with a spatula or wet hand by folding the edges into the center, usually about 30 folds. Flip the ball of dough in the bowl and cover.

Bulk Fermentation (2 to 4 hours)

1. Rest the dough at 74–80°F (23–27°C), with one or two sets of stretch and folds 15 to 30 minutes apart until the dough holds its ball shape somewhat and gives a small windowpane.

2. Let it rise until the dough is airy and puffy, double in size, and becoming jiggly (2 to 4 hours total time after mixing in the levain).

Shape the Loaf (20 minutes)

1. Turn the dough out onto a lightly misted counter and pre-shape: fold the dough into thirds, like a letter, fold in the other two sides, and flip the ball. Rotate it to tighten the skin. Let it rest 15 minutes until it relaxes. It will seem fairly sticky.

2. Prepare a wide bowl, plate, or platter sprinkled generously with rolled oats.

3. Shape into a boule or batard and tighten the skin well, watching closely for tearing. Mist the top and sides of the loaf lightly if the loaf has a lot of flour on it so the oats will stick.

4. Roll the loaf top side down onto the rolled oats until they adhere, then lift the loaf and place it in a well-floured proofing basket. Sprinkle rolled oats or flour around the edges where the dough tends to stick as it rises and lay parchment paper on top.

Proof the Loaf (1 to 12 hours)

Cover and let it rest at 74–80°F (23–27°C) until risen and jiggly like a water balloon (1 to 3 hours). Or retard the loaf in the refrigerator for 4 to 12 hours.

DAY 2 OR 3

Bake the Loaf (1½ to 2 hours)

1. Preheat the oven to 475°F (246°C) 30 to 60 minutes before baking, and prepare your baking and steam setup.

2. Turn out the loaf and score it using kitchen shears to cut through the oats.

3. Mist the loaf generously (optional) and transfer to the oven using your chosen baking and steam setup.

If baking with a cover: Cover and bake 20 minutes for the steam phase, then uncover and continue baking until the interior is 207°F (97°C) and the crust browns to your liking, usually 35 to 50 minutes total baking time.

If baking with a steam pan: Place the loaf in the oven, then pour 1½ cups (355 milliliters) of water into the steam pan. Bake 20 minutes without opening the oven for the steam phase, then continue baking until the interior is 207°F (97°C) and the crust browns to your liking, usually 35 to 50 minutes total baking time.

4. Cure in the cooling oven with the door ajar for 5 to 10 minutes. Cool 1 to 2 hours on a rack.

American Beauty

This sweet potato and pecan bread with cranberries screams fall. Because pecans, sweet potatoes, and cranberries are all native to the American continents, this bread is perfect for celebrating Thanksgiving. It's simply delightful with a triple cream brie. The sweet potato aroma is subtle while the pecans and the cranberries steal the show in this beautiful, fall-colors loaf.

The sweet potato definitely makes this a sticky, soft dough that's a little bit challenging to handle, but the bright orange hue of the dough will cheer you up.

MAKES ONE LOAF

TOTAL FLOUR: 450 GRAMS (1 POUND)
PRE-FERMENTED FLOUR: 17%
HYDRATION: 75%
WHOLE GRAIN: 0%
DOUGH: 1293 GRAMS (2.9 POUNDS)

DAY 1

Build the Levain (10 minutes)

1. In a small bowl or pint jar, mix together:

   30 grams starter that floats (2 tablespoons)

   60 grams water (¼ cup)

   60 grams all-purpose flour (½ cup)

2. Cover and let it rest at 60–70°F (16–21°C) for 8 to 12 hours until it is risen and bubbly and can float in water.

DAY 1 OR 2

Roast the Sweet Potatoes (1 to 1½ hours)

about 340 grams whole sweet potatoes (about ¾ pound)

1. Preheat oven to 400°F (204°C).
2. Pierce each potato several times with a fork to keep them from exploding in the oven.
3. Place them directly on a center rack in the oven, and place a sheet of foil on the bottom of the oven to catch any drips.
4. Roast until the skin is browned, you can smell the sugars caramelizing, and the flesh is very soft, 30 to 45 minutes, depending on the size of the sweet potatoes.
5. Cool, then remove the skin and mash well, but do not add any liquid. The potatoes can be made in advance: cover and refrigerate, warming to room temperature before using.

continued

DAY 2

Mix the Dough (10 minutes)

1. Measure the flour into a large heavy bowl and set aside:

375 grams bread flour (3⅛ cups)

2. Whisk together in a medium bowl until uniform:

264 grams water (1⅛ cups) at 90°F (32°C)

225 grams of the prepared sweet potato (scant 1 cup)

150 grams levain (all)

3. Mix the wet ingredients into the flour, combining completely. The dough will feel very sticky and wet.

Autolyse (30 to 60 minutes)

1. Rest the dough for 30 to 60 minutes until it has relaxed in the bowl.

2. Add:

9 grams salt (1½ teaspoons)

150 grams pecans (1½ cups)

120 grams dried sweetened cranberries (¾ cup)

3. Sprinkle over the dough and mix them in by poking and cutting them into the dough with a wet spatula, then folding the edges of the dough over the center with a spatula until the dough resists folding across itself and the pecans and cranberries are well distributed in the dough, about 30 folds. Flip the ball of dough in the bowl and cover.

Bulk Fermentation (3 to 5 hours)

1. Rest the dough at 74–80°F (23–27°C), with two or three sets of stretch and folds 15 to 30 minutes apart until the dough holds its ball shape somewhat.

2. Let it rise until the dough is airy and puffy, double or triple in size and jiggly (3 to 5 hours total time after mixing in the starter).

Shape the Loaf (20 minutes)

1. Handle the dough minimally and gently during shaping to achieve a light, lofty crumb. Turn the dough out onto a lightly misted counter, pre-shape, and rest 10 to 15 minutes until it relaxes.

2. Shape into a batard or boule and gently tighten the skin, being careful not to tear the skin. Poke in any nuts or cranberries that pop through the dough. Flour the top generously.

3. Place the loaf seam side up in a well-floured proofing basket. Sprinkle it with flour, especially around the edges. Lay parchment paper on top.

Proof the Loaf (1 to 18 hours)

Cover and retard the loaf in the refrigerator for 4 to 18 hours. Or let it rest at 74–80°F (23–27°C) until risen and jiggly like a water balloon (1 to 3 hours).

DAY 2 OR 3

Bake the Loaf

1. Preheat the oven to 475°F (246°C) 30 to 60 minutes before baking, and prepare your baking and steam setup.

2. Turn out the loaf, brush off excess flour, and lightly mist it.

3. Score the loaf and transfer to the oven using your chosen baking and steam setup.

If baking with a cover: Cover and bake 25 minutes for the steam phase, then uncover and continue baking until the interior is 207°F (97°C) and the crust browns to your liking, usually 35 to 50 minutes total baking time.

If baking with a steam pan: Place the loaf in the oven, then pour 1½ cups (355 milliliters) of water into the steam pan. Bake 25 minutes without opening the oven for the steam phase, then continue baking until the interior is 207°F (97°C) and the crust browns to your liking, usually 35 to 50 minutes total baking time.

4. Cure in the cooling oven with the door ajar for 5 to 10 minutes. Cool 1 to 2 hours on a rack.

Malted Barley Bread

For the beer and scotch lovers among us, this bread features sprouted barley baked right into the bread. When barley is soaked in water, the baby root emerges in the first day, and the baby shoot follows soon after. Along with these obvious changes, a delightful natural chemistry is triggered that results in maltose, a most intoxicating and fragrant sugar. This bread makes a fine accompaniment to a beef or lamb stew. Or try this bread with your favorite brew or single malt.

Be sure to use high-quality fresh, whole, hulled barley (not pearled barley) or it may not sprout. Expect faster than usual rising. The reduction in the oven temperature after steaming prevents the sugars in the crust from over-browning before the interior is baked.

MAKES ONE LOAF

TOTAL FLOUR: 450 GRAMS + 100 GRAMS BARLEY (1 POUND PLUS 0.2 POUND)

PRE-FERMENTED FLOUR: 17%

HYDRATION: 73%

WHOLE GRAIN FLOUR: 20%

TOTAL WHOLE GRAIN: 35%

DOUGH: ABOUT 1000 GRAMS (2.2 POUNDS)

A FEW DAYS AHEAD

Soak the Barley (5 minutes)

1. In a quart bowl or jar, mix together:

100 grams whole barley (½ cup)

about 350 grams water (1½ cups)

2. Rinse and drain the grains, then cover with the water. Cover and let it rest on the counter overnight.

3. Drain the grains, rinse gently, and drain completely so no grains are sitting in water. Cover with a clean cloth and let them rest for 8 to 12 hours.

4. Repeat the rinse-and-drain step every 8 to 12 hours until you see a little white root growing. For maximum malting activity, continue until the grain tastes sweet when you bite it, up to 3 days total.

DAY 1

Build the Levain (10 minutes)

1. In a small bowl or pint jar, mix together:

30 grams starter that floats (2 tablespoons)

60 grams water (¼ cup)

60 grams whole wheat flour (½ cup)

2. Cover and let it rest at 60–70°F (16–21°C) for 8 to 12 hours until it is risen and bubbly and can float in water.

continued

DAY 2

Mix Dough (15 minutes)

1. Mix the flours in a large heavy bowl and set aside:

285 grams bread flour (2⅜ cups)

90 grams whole wheat flour (¾ cup)

2. Whisk together in the levain bowl or jar until uniform:

150 grams levain (all)

254 grams water (1 cup plus 1 tablespoon) at 90°F (32°C)

3. Mix the wet ingredients into the flour, combining completely. The dough should feel a little firm but sticky.

Autolyse (30 to 60 minutes)

1. Rest the dough for 30 to 60 minutes at 74–80°F (23–27°C) until the dough is very relaxed in the bowl to help with incorporating the grains.

2. Rinse the grains under running water and leave to drain completely, blotting the strainer with a towel just before use. Roughly chop the grains with a chef's knife so all the grains are cut, releasing their enzymes.

3. Sprinkle onto the surface of the dough:

9 grams salt (1½ teaspoons)

all the chopped barley

4. Poke and pinch the salt and grains into the dough. Thoroughly mix the dough in the bowl by folding the edges into the center until the dough is well mixed and forms a springy ball, usually 30 folds. Flip the ball of dough in the bowl and cover.

Bulk Fermentation (2 to 4 hours)

1. Rest the dough at 74–80°F (23–27°C), with one or two sets of stretch and folds 15 to 30 minutes apart until the dough holds its ball shape somewhat and gives a small windowpane.
2. Let it rise until the dough is airy and puffy, double in size, and becoming jiggly (2 to 4 hours total time after mixing in the levain).

Shape the Loaf (25 minutes)

1. Turn the dough out onto a lightly misted counter and pre-shape: fold the dough into thirds, like a letter, fold in the other two sides, and flip the ball. Rotate it to tighten the skin. Let it rest 10 to 15 minutes until it relaxes.
2. Shape into a boule or batard and gently tighten the skin, watching closely for tearing. Flour the top generously.
3. Place the loaf seam side up in a well-floured proofing basket. Sprinkle it with flour, especially around the edges. Lay parchment paper on top.

Proof the Loaf (30 minutes to 12 hours)

Cover and let it rest at 74–80°F (23–27°C) until risen and jiggly like a water balloon (30 to 90 minutes). Or retard the loaf in the refrigerator for 4 to 12 hours.

DAY 2 OR 3

Bake the Loaf (1½ to 2 hours)

1. Preheat the oven to 475°F (246°C) 30 to 60 minutes before baking, and prepare your baking and steam setup.
2. Turn out the loaf and score it. Mist the loaf (optional) and transfer to the oven using your chosen baking and steam setup.

If baking with a cover: Cover and bake 20 minutes for the steam phase, then uncover and continue baking until the interior is 207°F (97°C) and the crust browns to your liking, usually 35 to 50 minutes total baking time.

If baking with a steam pan: Place the loaf in the oven, then pour 1½ cups (355 milliliters) of water into the steam pan. Bake 20 minutes without opening the oven for the steam phase, then continue baking until the interior is 207°F (97°C) and the crust browns to your liking, usually 35 to 50 minutes total baking time.

3. Cure in the cooling oven with the door ajar for 5 to 10 minutes. Cool 1 to 2 hours on a wire rack.

Rye Revelation

This whole grain rye bread has an amazing floral aroma, a deeply lactic but not overly sour flavor, and a creamy, chewy texture. It's all brought about by treating this quirky grain the way it needs to be treated, rather than shoehorning it into our wheat dough routine. As a bonus, this bread is very easy to make. Use very fresh flour to get all the subtle rye aromas in the bread.

This all-rye bread is unusual in every respect compared to the wheat breads in this cookbook. The recipe calls for a larger levain to get the dough off to a fast start for a short bulk fermentation. We dispense with most handling of this abominably sticky dough to avoid the gumminess that would result from breaking down the delicate rye pentosan structure. We let the loaf score itself, and we'll age the bread for a day before cutting into the loaf.

MAKES ONE LOAF

TOTAL FLOUR: 450 GRAMS (1 POUND)

PRE-FERMENTED FLOUR: 25%

HYDRATION: 80%

WHOLE GRAIN: 97%

DOUGH: 819 GRAMS (1.8 POUNDS)

DAY 1

Build the Levain (10 minutes)

1. In a small bowl or pint jar, mix together:

> **30 grams starter that floats (2 tablespoons)**
>
> **100 grams water (3/8 cup plus 1 tablespoon)**
>
> **100 grams whole rye flour (7/8 cup)**

2. Cover and let it rest at 60–70°F (16–21°C) for 8 to 12 hours until it is risen and bubbly and can float in water.

DAY 2

Mix the Dough (10 minutes)

1. Measure the flour into a large heavy bowl and set aside:

> **335 grams whole rye flour (2¾ cups)**

2. Whisk together in the levain bowl or jar until uniform:

> **245 grams water at 90°F (32°C)**
>
> **230 grams levain (all)**
>
> **9 grams salt (1½ teaspoons)**

3. Mix the wet ingredients into the rye flour, combining completely, then stop mixing. The dough will seem extremely sticky and shaggy, resembling cookie dough.

4. Let the dough rest for 10 minutes, then fold the edges of the dough over the center with a spatula until the dough is well mixed and smooth, 15 to 20 folds. The dough will seem clay-like, stiff, gummy, and brittle and will not be stretchy at all. Pat the dough into a ball in the bowl and cover.

Bulk Fermentation (2 to 12 hours)

Let the dough rest undisturbed at 74–80°F (23–27°C) until it is puffy and double in size—cracks may appear in the surface—(2 to 3 hours total time after mixing in the levain). Or rest the dough in the refrigerator for 8 to 12 hours.

Shape the Loaf (10 minutes)

1. Turn the dough out onto a lightly floured counter and very gently fold in the four sides, pinch them together, and flip the ball.
2. Mold gently into a boule by cupping your hands around the dough. Flour the top generously and allow the loaf to rest for 5 minutes to seal the seam.
3. Place the loaf seam side up in a very well-floured and lined proofing basket. Sprinkle it with flour, especially around the edges. Lay parchment paper on top.

Proof the Loaf (1 to 3 hours)

Cover and let it rest at 74–80°F (23–27°C) until risen 1½ to 2 times the original size and bubbles and cracks are visible in the surface (1 to 3 hours).

Bake the Loaf (1½ to 2 hours)

1. Preheat the oven to 450°F (232°C) 30 to 60 minutes before baking, and prepare your baking and steam setup.
2. Turn out the loaf and let it rest for about 15 minutes until several cracks have formed in the loaf.
3. Transfer to the oven using your chosen baking and steam setup.

If baking with a cover: Cover and bake 20 minutes for the steam phase, then uncover and continue baking until the interior is 207°F (97°C) and the crust browns to your liking, usually 40 to 55 minutes total baking time.

If baking with a steam pan: Place the loaf in the oven, then pour 1½ cups (355 milliliters) of water into the steam pan. Bake 20 minutes without opening the oven for the steam phase, then continue baking until the interior is 207°F (97°C) and the crust browns to your liking, usually 40 to 55 minutes total baking time.

4. Cure in the cooling oven with the door ajar for 5 to 10 minutes. Cool 1 to 2 hours on a rack. For the best flavor and texture, let the loaf rest for a day before cutting into it.

Sunflower Spelt Boule

A light and springy open crumb is encased in oven-toasted sunflower seeds and a crackling-crisp crust in this 50 percent whole grain spelt bread. The nutty flavor of spelt shines.

This is a very soft and stretchy dough that will reward your gentle handling with a light, open crumb. Spelt makes the dough very relaxed, so this loaf will spread into a wide round while baking. If you want a taller boule as in the photograph, use a baking vessel that will support the dough as it rises in the oven, such as a 3-quart (3-liter) casserole dish or Dutch oven. You can also make this bread with emmer, einkorn, khorasan, or durum in place of spelt.

MAKES ONE LOAF

TOTAL FLOUR: 450 GRAMS (1 POUND)

PRE-FERMENTED FLOUR: 17%

HYDRATION: 85%

WHOLE GRAIN: 50%

DOUGH: 845 GRAMS (1.9 POUNDS)

DAY 1

Build the Levain (10 minutes)

1. In a small bowl or pint jar, mix together:

30 grams starter that floats (2 tablespoons)

60 grams water (¼ cup)

60 grams whole wheat flour (½ cup)

2. Cover and let it rest at 60–70°F (16–21°C) for 8 to 12 hours until it is risen and bubbly and can float in water.

DAY 2

Mix the Dough (10 minutes)

1. Mix the flour in a large heavy bowl and set aside:

135 grams spelt flour (1⅛ cups)

30 grams whole wheat flour (¼ cup)

210 grams bread flour (1¾ cups)

2. Whisk together in the levain bowl or jar until uniform:

308 grams water (1¼ cups) at 90°F (32°C)

150 grams ready levain (all)

3. Mix the wet ingredients into the flour, combining completely. The dough will seem very wet and shaggy.

Autolyse (30 to 60 minutes)

1. Rest the dough for 30 to 60 minutes until the dough has relaxed in the bowl.

2. Add:

9 grams salt (1½ teaspoons)

3. Sprinkle the salt over the dough and mix it in by poking and cutting it into the dough with a wet spatula, then folding the edges of the dough over the center with a spatula until the dough resists folding across itself, about 20 folds. Flip the ball of dough in the bowl and cover.

Bulk Fermentation (3 to 5 hours)

1. Rest the dough at 74–80°F (23–27°C), with two or three sets of stretch and folds 15 to 30 minutes apart until the dough holds its ball shape somewhat and makes a small windowpane.

2. Let the dough rest until it is risen and seems light, with bubbles throughout, the smell of flour is gone and it smells more like bread, the dough jiggles in its bowl when you shake it, usually 4 hours total time after mixing in the levain.

continued

Shape the Loaf (20 minutes)

1. Handle the dough gently to avoid tearing the skin or losing loft. Turn the dough out onto a lightly misted counter and pre-shape: fold the dough into thirds, like a letter, fold in the other two sides, and flip the ball. Rotate it to tighten the skin. Let it rest 10 to 15 minutes until it relaxes.

2. Shape the dough into a boule. It will be hard to get it to stand tall because spelt flour makes the dough want to spread out and not stay in place when you fold it over itself.

3. Tighten the skin and let it sit a couple of minutes to seal the bottom.

4. Sprinkle most of ¼ cup (35 grams) raw, unsalted sunflower seed meats onto a sheet of waxed paper on the counter.

5. Oil the top and sides of the loaf and place it, top side down, on the seeds on the waxed paper. Lift the dough and waxed paper into the proofing basket.

6. Sprinkle the remaining seeds around edges where dough tends to stick as it rises and lay parchment paper on top.

Proof the Loaf (1 to 12 hours)

Cover and retard the loaf in the refrigerator for 4 to 12 hours. Or let it rest at 74–80°F (23–27°C) until risen and jiggly like a water balloon (1 to 3 hours).

DAY 2 OR 3

Bake the Loaf

1. Preheat the oven to 475°F (246°C) 30 to 60 minutes before baking, and prepare your baking and steam setup (covered baker recommended).

2. Turn out the loaf and score it. For a sun pattern, make a circle with a dotted line surrounded by radial lines.

3. Mist the loaf generously (optional) and transfer to the oven using your chosen baking and steam setup.

If baking with a cover: Cover and bake 20 minutes for the steam phase, then uncover and continue baking until the interior is 207°F (97°C) and the crust browns to your liking, usually 35 to 50 minutes total baking time.

If baking with a steam pan: Place the loaf in the oven, then pour 1½ cups (355 milliliters) of water into the steam pan. Bake 20 minutes without opening the oven for the steam phase, then continue baking until the interior is 207°F (97°C) and the crust browns to your liking, usually 35 to 50 minutes total baking time.

4. Cure in the cooling oven with the door ajar for 5 to 10 minutes. Cool 1 to 2 hours on a rack.

Golden Gate Gem

A golden, crispy crust and a tart, chewy crumb are the hallmarks of the historic and world-famous San Francisco sourdough bread. In this loaf, we take sour to the max over a three-day journey. The large levain and warm fermentations build up the population of lactic acid bacteria at their favorite temperatures. Then a long, cold retard seals the deal, giving those bacteria ample opportunity to churn out acetic acid for the tang expected in a San Francisco sourdough. Finally, we perfect the blistered, crispy golden crust during the bake with a water mist and ample steam. Heaven!

Steam is critical for the crispy crust, so a cover or Dutch oven is recommended unless your oven has a steam function. For an all-white sourdough, you can substitute white flour for the whole wheat and use only 165 grams water when mixing the dough (½ cup plus 3 tablespoons).

MAKES ONE LOAF

TOTAL FLOUR: 450 GRAMS (1 POUND)

PRE-FERMENTED FLOUR: 33.3%

HYDRATION: 72%

WHOLE GRAIN: 15%

DOUGH: 783 GRAMS (1.7 POUNDS)

DAY 1

Build the Levain (10 minutes)

1. In a small bowl or pint jar, mix together:

> **10 grams starter (2 teaspoons; no need to have fed it recently for this recipe)**
>
> **145 grams all-purpose flour (1¼ cups)**
>
> **145 grams water (⅝ cup) at 90°F (32°C)**

2. Leave at 80–85°F (27–29°C) until it has doubled in volume, then refrigerate for 8 to 24 hours.

DAY 2

Mix the Dough (10 minutes)

1. Measure the flour into a large heavy bowl:

> **232 grams bread flour (scant 2 cups)**
>
> **68 grams whole wheat flour (½ cup plus 1 tablespoon)**

2. Whisk together in the levain bowl or jar until uniform:

> **174 grams water (¾ cup) at 90°F (32°C)**
>
> **300 grams levain (all)**

continued

Autolyse (30 to 60 minutes)

1. Mix until all the flour is wet but do not knead. Cover and let the dough rest for 30 minutes at 74–80°F (23–27°C).
2. Add:

9 grams salt (1½ teaspoons)

3. Sprinkle the salt over the wet dough and mix it in by poking and cutting it into the dough with a wet spatula, then folding the edges of the dough over the center with a spatula until the dough resists folding across itself, usually about 20 times. The dough will seem sticky but firm. Flip the ball of dough and cover.

Bulk Fermentation (6 to 15 hours)

1. Rest the dough at 74–80°F (23–27°C), with two or three sets of stretch and folds 15 to 30 minutes apart until the dough holds its shape well after folding and passes the windowpane test.
2. Continue resting until the dough has almost doubled (2 to 3 hours total after adding the levain).
3. Stretch and fold the dough once around, then move the dough to the refrigerator for 4 to 12 hours.

DAY 3

Shape the Loaf (25 minutes)

1. Handle the dough minimally and gently during shaping to achieve a light, lofty crumb. Turn the dough out onto a lightly misted counter and pre-shape: fold the dough into thirds, like a letter, fold in the other two sides, and flip the ball. Rotate it to tighten the skin. Let it rest 15 minutes until it relaxes.
2. Shape into a boule or batard and gently tighten the skin, watching closely for tearing. Flour the top generously.
3. Place the loaf seam side up in a well-floured proofing basket. Sprinkle with flour, especially around the edges. Lay parchment paper on top.

Proof the Loaf (30 to 90 minutes)

Cover the loaf to prevent drying and let it rest at room temperature until risen and jiggly like a water balloon, 30 to 90 minutes.

continued

Bake the Loaf

1. Preheat the oven to 475°F (246°C) 30 to 60 minutes before baking, and prepare your baking and steam setup. A covered baking setup is ideal for this bread.
2. Turn out the loaf, brush away loose flour, and score it.
3. Mist the loaf generously, as well as the inside of the cover, and transfer to the oven using your chosen baking and steam setup.

If baking with a cover: Cover and bake 20 minutes for the steam phase, then uncover and continue baking until the interior is 207°F (97°C) and the crust browns to your liking, usually 35 to 50 minutes total baking time.

If baking with a steam pan: Place the loaf in the oven, then pour 1½ cups (355 milliliters) of water into the steam pan. Bake 20 minutes without opening the oven for the steam phase, then continue baking until the interior is 207°F (97°C) and the crust browns to your liking, usually 35 to 50 minutes total baking time.

4. Cure in the cooling oven with the door ajar for 5 to 10 minutes. Cool 1 to 2 hours on a rack.

Demi-Miche

A miche is traditionally a very large round whole grain sourdough loaf, typically using 1 to 5 kilograms of rustic, stone-ground flour. The loaf was meant to last for many days, as families had to take turns at the communal oven, back in the day. Nowadays, large miches, like the famed miche Poilâne, are often cut into pieces and sold or purchased whole for large gatherings. A good miche has a complex flavor that develops over two or three days, due to the long fermentation and complex, rustic flours. This bread is an excellent choice for a tartine.

This recipe makes a normal-sized loaf, but feel free to double or triple the recipe to make a bigger loaf. Just be sure to use a bigger baking setup and expect to add 15 to 25 minutes to the baking time.

MAKES ONE LOAF

TOTAL FLOUR: 450 GRAMS (1 POUND)

PRE-FERMENTED FLOUR: 25%

HYDRATION: 75%

WHOLE GRAIN: 70%

DOUGH: 787 GRAMS (1.7 POUNDS)

DAY 1

Build the Levain (10 minutes)

1. In a small bowl or pint jar, mix together:

   30 grams starter that floats (2 tablespoons)

   98 grams water (½ cup minus 1 tablespoon)

   45 grams whole rye flour (⅜ cup)

   52 grams whole wheat flour (½ cup minus 1 tablespoon)

2. Cover and let it rest at 60–70°F (16–21°C) for 8 to 12 hours until it is risen and bubbly and can float in water.

continued

DAY 2

Mix the Dough (10 minutes)

1. Mix the flour in a large heavy bowl and set aside:

127 grams whole wheat flour (1 cup plus 1 tablespoon)

120 grams bread flour (1 cup)

90 grams whole spelt flour (¾ cup)

2. Whisk together in the levain bowl or jar until uniform:

225 grams water (1 cup minus 1 tablespoon) at 90°F (32°C)

225 grams levain (all)

3. Mix the wet ingredients into the flour, combining completely.

Autolyse (30 to 60 minutes)

1. Rest the dough for 30 to 60 minutes until it has relaxed in the bowl.

2. Add:

9 grams salt (1½ teaspoons)

3. Sprinkle the salt over the wet dough and mix it in by poking and cutting it into the dough with a wet spatula, then folding the edges of the dough over the center with a spatula until the dough resists folding across itself, about 20 folds. Flip the ball of dough in the bowl and cover.

Bulk Fermentation (10 to 14 hours)

1. Rest the dough at 74–80°F (23–27°C), with two or three sets of stretch and folds 15 to 30 minutes apart until the dough holds its ball shape somewhat and makes a nice windowpane. Rest until the dough begins to rise and become puffy (2 to 3 hours after mixing in the levain). Then transfer the dough to the refrigerator for 8 to 12 hours.

2. Bring the dough out of the refrigerator. It should be risen to just about double in size, but if it isn't, you can allow it to rise a bit more at room temperature before turning it out.

continued

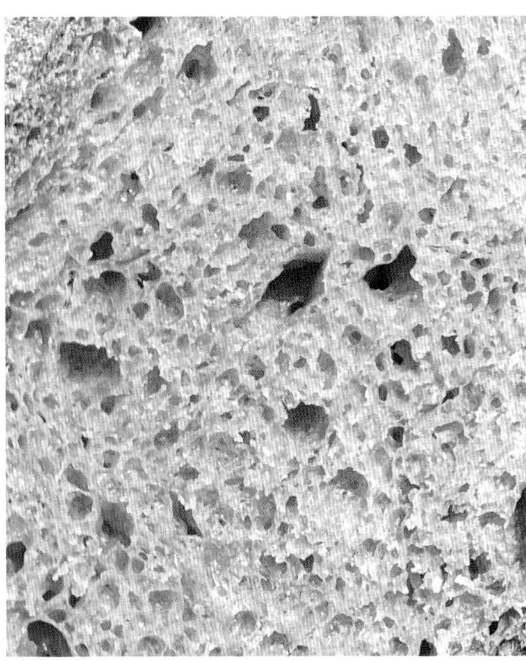

Shape the Loaf (20 minutes)

1. Turn the dough out onto a lightly misted counter. Handle the dough slowly and gently during shaping to achieve a light crumb. Pre-shape: fold the dough into thirds, like a letter, fold in the other two sides, and flip the ball. Rotate it to tighten the skin. Let it rest 10 to 15 minutes or longer until it relaxes and warms up.

2. Shape into a boule and gently tighten the skin, being careful not to tear the skin. Flour the top generously.

3. Place the loaf seam side up in a well-floured proofing basket. Sprinkle it with flour, especially around the edges. Lay parchment paper on top. For a larger miche, use a lined large bowl or colander with room for the loaf to rise. If you don't have anything big enough, place the loaf seam side down on a parchment paper–lined baking sheet or peel, dust with flour, and lightly cover with parchment paper then a damp tea towel or a plastic bag.

Proof the Loaf (1 to 3 hours)

Cover and let it rest at 74–80°F (23–27°C) until risen and jiggly like a water balloon (1 to 3 hours).

Bake the Loaf

1. Preheat the oven to 475°F (246°C) 30 to 60 minutes before baking, and prepare your baking and steam set-up.

2. Turn out the loaf and score it several times all around in a shallow manner.

3. Mist the loaf generously (optional) and transfer to the oven using your chosen baking and steam setup.

If baking with a cover: Cover and bake 20 minutes for the steam phase, then uncover and continue baking until the interior is 207°F (97°C) and the crust browns to your liking, usually 35 to 50 minutes total baking time.

If baking with a steam pan: Place the loaf in the oven, then pour 1½ cups (355 milliliters) of water into the steam pan. Bake 20 minutes without opening the oven for the steam phase, then continue baking until the interior is 207°F (97°C) and the crust browns to your liking, usually 35 to 50 minutes total baking time.

4. Cure in the cooling oven with the door ajar for 5 to 10 minutes. Cool 1 to 2 hours on a rack.

Kiddo's Milk Bread

If life throws you a person to bake for who craves a squarish piece of white bread with a buttery, tender crust, this recipe will get you there in grand style. Using a fresh young levain, we keep the sourness toned down, and the milk brings a velvet creaminess to the crumb. A generous brush of butter adds depth and flavor to this delicate bread.

The milk makes the dough feel rather tacky, but after the stretches, it should become easy to manage for shaping.

MAKES ONE LOAF

TOTAL FLOUR: 450 GRAMS (1 POUND)

PRE-FERMENTED FLOUR: 17%

HYDRATION: 80%

WHOLE GRAIN: 0%

DOUGH: 819 GRAMS (1.8 POUNDS)

DAY 1

Build the Levain (10 minutes)

1. In a small bowl or pint jar, mix together:

30 grams starter that floats (2 tablespoons)

60 grams water (¼ cup)

60 grams all-purpose flour (½ cup)

2. Cover and let it rest at 60–70°F (16–21°C) for 6 to 10 hours until it is just doubled and is bubbly and can float in water.

DAY 2

Mix the Dough (10 minutes)

1. Measure the flour into a large heavy bowl and set aside:

375 grams bread flour (3⅛ cups)

2. Whisk together in the levain bowl or jar until uniform:

285 grams whole milk (1 cup plus 3 tablespoons) at 90°F (32°C)

150 grams levain (all)

3. Mix the wet ingredients into the flour, combining completely. The dough will seem very sticky but should be able to form a loose ball.

Autolyse (30 to 60 minutes)

1. Rest the dough for 30 to 60 minutes until it has relaxed in the bowl.

2. Add:

9 grams salt (1½ teaspoons)

continued

3. Sprinkle the salt over the dough and mix it in by poking and cutting it into the dough with a wet spatula, then folding the edges of the dough over the center with a spatula until the dough resists folding across itself, about 20 folds. Flip the ball of dough in the bowl and cover.

Bulk Fermentation (3 to 5 hours)

1. Rest the dough at 74–80°F (23–27°C), with two or three sets of stretch and folds 15 to 30 minutes apart until the dough holds its ball shape, becomes less sticky, and makes a nice windowpane.

2. Let it rise until the dough is airy and puffy, double or triple in size, jiggly, and smells like bread (3 to 5 hours total time after mixing in the levain).

Shape the Loaf (20 minutes)

1. Handle the dough minimally and gently during shaping to achieve a light, lofty crumb. Turn the dough out onto a lightly misted counter and pre-shape: fold the dough into thirds, like a letter, fold in the other two sides, and flip the ball. Rotate it to tighten the skin. Let it rest 10 to 15 minutes until it relaxes.

2. Shape into a batard and gently tighten the skin, being careful not to tear the skin.

3. Place seam side down in a generously buttered bread pan and brush or gently rub softened butter on top.

Proof the Loaf (1 to 24 hours)

Cover and retard the loaf in the refrigerator for 4 to 24 hours. Or let it rest at 74–80°F (23–27°C) until risen and jiggly like a water balloon (1 to 3 hours).

Bake the Loaf

1. Preheat the oven to 400°F (204°C) and prepare a steam pan.

2. Score the loaf.

3. Transfer the pan to the oven and pour 1½ cups (355 milliliters) of water into the steam pan and close the oven door.

4. Bake 25 to 35 minutes until the interior is at least 195°F (91°C) and the crust browns to your liking, usually a medium golden brown.

5. Cool for 5 to 10 minutes in the pan, then turn out the loaf and cool 1 to 2 hours on a rack.

CHAPTER 9

Let Nature Take Its Course: Slow Fermentation Recipes

These slow recipes are great for many reasons, mostly because they make fantastic bread, but also because they demand very little effort on the part of the baker. Mix up all the ingredients in 10 minutes, then forget about it for 10 to 18 hours. The rest of the recipe is like any other in this book.

If you've been paying attention, you've noticed that there are not many recipes in this book that call for any kneading. Instead, we do easy stretch and folds and spatula folds to work the dough gently. There is even less working of the dough in these slow recipes, and so they are similar to the no-knead recipes popularized over the past couple of decades.

Instead of working the dough to develop gluten, we slow down the bulk fermentation and use a high hydration level so that the gluten network can develop spontaneously. We use a cooler temperature for the water and bulk fermentation to slow the growth of the sourdough yeast and bacteria so they don't run out of food or over-rise while the gluten network slowly forms. We also mix a lot less pre-fermented flour into the dough, so the population of sourdough organisms takes a longer time to get to the peak. Fermenting at a cool temperature and using a small amount of starter are critical to success.

The recipes in this chapter cover a variety of breads and are arranged with the easier recipes first. Toward the end of the chapter are some breads that will be easier to make successfully once you are pretty comfortable with the techniques described in this book and that are noted in the recipe descriptions.

Pros and Cons of a Slow Fermentation Recipe

No-knead recipes can be developed for many kinds of bread, but one place they really shine is for doughs with gluten networks that are sensitive to over-manipulation. For example, certain grains have small amounts of gluten, such as rye, and others have gluten that forms delicate bonds, such as spelt or durum. Additionally, doughs with a lot of whole grain benefit from the long period of sitting undisturbed, during which time the bran absorbs water and softens. The dough is then manipulated for shaping the loaf. This two-step process protects the gluten network. You might be surprised at how light a whole grain bread can be if you prevent damage to its gluten network by simply leaving it alone.

Because these recipes are less work, you may be wondering why this isn't the beginner chapter. The main reason is that while dough is left untended for long periods, the baker is not there to observe the progression of changes. You, the baker, must therefore have a good feel for where a bowl of dough is in its fermentation, and you must have a good sense for what has happened while you weren't there. This isn't an issue for an experienced baker. But novices may not have enough experience to know when the dough is ready, and the hands-off bulk fermentation doesn't provide much of a chance for them to learn since they aren't there to watch it happen. Here's a list of the considerations when choosing a slow fermentation recipe versus a levain recipe.

PROS:

- No hands-on time or effort required once the dough is mixed until the end of the bulk fermentation.
- There's a bigger "ready" window for the bulk fermentation because it is happening more slowly.
- Mixing everything at once leads to less cleanup.
- It's easier to fit these recipes into the middle of a busy schedule.
- You can make them with unrefreshed starter, though a recently refreshed starter is always a safer bet.
- Long fermentation times lead to greater flavor development as well as digestibility.

CAVEATS:

- No hands-on time means you're not there to see the bulk fermentation and to learn about what is happening, or to enjoy the experience of handling the dough.
- Out of sight, out of mind—you may forget to check on things and accidentally let the fermentation go too far.
- It can be hard to tell when a fermentation is ready if you just look at a snapshot in time rather than compare how the dough has changed over time.
- No-knead recipes can seem deceptively simple. There's nothing to do, so you can't mess it up! However, being able to judge whether your bulk fermentation has under-proofed, over-proofed, or is just ready is the key to success. In the absence of this knowledge, strict adherence to the

times, temperatures, and measurements will lead to a great loaf.

- If your kitchen is warmer than 70°F (21°C), you will need to engineer a way to keep the dough cool.

You may also be wondering why anyone bothers working the dough at all? As you'll come to appreciate from these recipes, developing the gluten via folding or kneading the dough can really speed up the process, and it also gives the baker more control over the amount of gluten development. For low-hydration bread recipes, in which spontaneous gluten development is hindered by lack of sufficient water, gluten formation by way of mechanical means can improve the bread significantly. That said, if you have enough lead time before you want the bread to be ready, a slow fermentation recipe is a great choice for convenience and the overall quality of the loaf.

Colder Dough Needs Longer Rising Time

In these recipes, we are using pretty cool temperatures for the bulk fermentation (compared to the levain recipes) in order to slow down fermentation. Remember that temperature has a powerful influence on the speed of fermentation. At 60°F (16°C) it will take many more hours for the dough to rise than at 70°F (21°C). That is why the bulk fermentation time is given as 10 to 18 hours for most of the recipes in this chapter. Just remember, the colder the temperature, the longer it will take for the dough to rise, so plan accordingly.

Using a Cold Retard in a Slow Fermentation

With the long bulk fermentation, an overnight retard may be too much, causing an over-fermented loaf that cannot rise in the oven. When an overnight retard is not recommended, I don't offer it as an option in the recipe. However, a cold retard improves flavor, quality of the crust, and flexibilty, and I encourage you to use a cold retard whenever you can in these recipes. Simply plan it out and read your dough to determine how far along its fermentation has gone if you want to retard your loaf. Note that it will take the dough at least a couple of hours to chill in the refrigerator, and during that time it is still fermenting at its cooler temperature.

In general, if you plan to retard the loaf, end bulk fermentation a little on the early side when the dough has just approached double its volume. Put the loaf directly in the refrigerator after forming it and keep it cold until turning it out to bake. Don't worry if it looks like it hasn't risen: It will spring in the oven if you have retarded it for the amount of time called for in the recipe. This is my favorite strategy because I love the crust and crumb of a richly fermented loaf, and I adore a bodacious oven spring.

If you plan to retard the bulk fermentation, before refrigerating allow the dough to begin rising so that it expands to at least 1¼ times the original volume and bubbles are clearly forming. Bring the dough out to warm (and to finish rising if necessary) before forming the loaf.

Fitting the Recipe to Your Schedule

With minimal hands-on demands, it is easy to fit a slow fermentation recipe to your preferred timeframe.

RHYTHM 1. The morning person: Mix the dough the first thing in the morning on Day 1. When it's ready in the evening, shape the loaf. Retard the loaf overnight in the refrigerator and bake it the following morning on Day 2.

RHYTHM 2. The night owl: Mix the dough in the evening on Day 1. When it's ready the next morning on Day 2, shape the loaf. Proof and bake right away or retard the loaf for a few hours in the refrigerator and bake later that day.

RHYTHM 3. The crazy busy person: Mix the dough when you can on Day 1. After a couple of hours at room temperature, move the dough to the refrigerator for 20 to 30 hours. Then on Day 2 let it come to room temperature, shape the loaf, proof it, and bake it.

Slow and Easy

It's amazing how delicious this bread is for so little trouble. Slow and Easy is similar to the Simple Boule, but it's even easier to make. Time and nature do most of the work while you do other things. You'll notice a richer flavor thanks to the extended fermentation time. If your kitchen is very cold, using 90°F (32°C) water to mix the dough will get things off to a good start. If your kitchen is warm, you will need to find a way to cool the dough to extend the fermentation or expect the dough to be ready much sooner. Success is easy, but perfecting the flavor hinges on ending the fermentation and proofing steps when the time is right. Luckily, the "ready" window is wide at these cooler temperatures.

MAKES ONE LOAF

TOTAL FLOUR: 450 GRAMS (1 POUND)

PRE-FERMENTED FLOUR: 3%

HYDRATION: 72%

WHOLE GRAIN: 0%

DOUGH: 783 GRAMS (1.7 POUNDS)

DAY 1

Mix the Dough (10 minutes)

1. Measure the flour into a large heavy bowl and set aside:

435 grams bread flour (3⅝ cups)

9 grams salt (1½ teaspoons)

2. Whisk the wet ingredients together in a small bowl until uniform:

309 grams water (1⅜ cups)

30 grams starter that floats (2 tablespoons)

3. Mix the wet ingredients into the flour, combining completely. The dry flour needs to be fully incorporated.

Bulk Fermentation (10 to 18 hours)

1. Rest the dough 10 to 60 minutes and then perform a set of stretch and folds so the dough forms a nice, firm ball and flip the ball.

2. Cover the dough and let it rest at 60–70° (16–21°C) for 10 to 18 hours until the dough is puffed up, very jiggly, and smells ready.

continued

DAY 1 OR 2

Shape the Loaf (15 minutes)

1. Turn the dough out gently onto a damp counter. It should seem lively, wet, and bubbly.
2. Pre-shape: fold the dough into thirds, like a letter, fold in the other two sides, and flip the ball. Rotate it to tighten the skin. Let it rest for 10 minutes.
3. Shape the dough into a ball or batard without deflating it and tighten the skin.
4. Sprinkle the entire top and sides of the loaf with a light coating of flour so it won't stick during proofing.
5. Place the loaf seam side up and in a well-floured proofing basket. Sprinkle it with flour, especially around the edges. Lay parchment paper on top.

Proof the Loaf (1 to 24 hours)

Cover and let it rest at room temperature for 1 to 3 hours until the loaf seems jiggly like a water balloon. Or retard the loaf in the refrigerator for 4 to 24 hours provided you ended the bulk fermentation early.

DAY 2 OR 3

Bake the Loaf (1½ to 2 hours)

1. Preheat the oven to 475°F (246°C) 30 to 60 minutes before baking, and prepare your baking and steam setup.
2. Turn out the loaf and dust off excess flour.
3. Score the loaf and mist generously (optional).
4. Transfer the parchment paper and loaf to the baking setup.

If baking with a cover: Cover and bake 20 minutes for the steam phase, then uncover and continue baking until the interior is 207°F (97°C) and the crust browns to your liking, usually 35 to 50 minutes total baking time.

If baking with a steam pan: Place the loaf in the oven, then pour 1½ cups (355 milliliters) of water into the steam pan. Bake 20 minutes without opening the oven for the steam phase, then continue baking until the interior is 207°F (97°C) and the crust browns to your liking, usually 35 to 50 minutes total baking time.

5. Cure in the cooling oven with the door ajar 5 to 10 minutes. Cool 1 to 2 hours on a rack.

Artisan White

In this loaf, a caramelized, blistered, crunchy crust encases a tender, chewy crumb that's studded with shiny open holes. The flavor is top-notch, full of complex subtle notes due to the long, slow fermentation and the high amount of water in the dough.

It is very tricky to handle the high-hydration dough the first couple of times (review Handling Sticky, High-Hydration Dough on page 27 before beginning), but it's so worth the trouble! If it's your first time, try dividing the dough into two mini loaves to make the shaping process easier. Expect a faster fermentation because of the higher hydration, but don't be afraid to let the dough really rise during bulk fermentation (or else include a long retard). A long, slow fermentation is the magic ingredient!

MAKES ONE LOAF

TOTAL FLOUR: 450 GRAMS (1 POUND)
PRE-FERMENTED FLOUR: 3%
HYDRATION: 80%
WHOLE GRAIN: 2%
DOUGH: 819 GRAMS (1.8 POUNDS)

DAY 1

Mix the Dough (10 minutes)

1. Mix the flour in a large heavy bowl and set aside:

426 grams bread flour (3½ cups)

9 grams diastatic barley malt flour (1 tablespoon)

9 grams salt (1½ teaspoons)

2. Whisk the wet ingredients together in a small bowl until uniform:

345 grams water (1½ cups)

30 grams starter that floats (2 tablespoons)

3. Mix the wet ingredients into the flour, combining completely. The dry flour needs to be fully incorporated. The dough will seem very shaggy and wet, but do not add more flour.

Bulk Fermentation (10 to 18 hours)

1. Cover the dough and let it rest at 60–70°F (16–21°C) for 30 to 60 minutes. Then fold the dough with a wet spatula until it forms a wet, loose ball and flip the ball. The dough should now feel soft and strong. If not, rest the dough for 15 to 30 minutes, then perform additional spatula folds to develop strength.

continued

2. Continue resting for 10 to 18 hours total until the dough is very puffed up to two or three times its original size, is jiggly, and smells ready, with many large bubbles visible on the surface. (If you plan on a loaf retard, end the bulk fermentation at two times its original size.)

DAY 1 OR 2

Shape the Loaf (15 to 20 minutes)

1. Turn the dough out gently onto a damp counter. It should seem lively, wet, and bubbly.

2. Pre-shape: fold the dough into thirds, like a letter, fold in the other two sides, and flip the ball. Rotate it to tighten the skin. Let it rest for 10 to 15 minutes. It will seem like a big water balloon.

3. Shape the dough into a ball or batard without deflating it and tighten the skin.

4. Sprinkle the entire top and sides of the loaf with a generous coating of flour.

5. Place the loaf seam side up in a well-floured proofing basket. A linen cloth liner and generous flour help to keep this wet dough from sticking.

6. Sprinkle it with flour, especially around the edges. Lay parchment paper on top.

Proof the Loaf: (30 minutes to 24 hours)

Cover and let it rest at room temperature for 30 to 90 minutes until the loaf has risen and is jiggly like a water balloon. Since we extended the bulk fermentation and retained most of the loft in the dough, a short loaf proof is sufficient and will help us get good oven spring. Or retard the loaf in the refrigerator for 4 to 24 hours, provided you ended the bulk fermentation early.

DAY 2 OR 3

Bake the Loaf (1½ to 2 hours)

1. Preheat the oven to 500°F (260°C) 30 to 60 minutes before baking, and prepare your baking and steam setup.

2. Turn out the loaf and dust off excess flour. The loaf may spread out quite a lot.

3. Score the loaf and mist generously (optional).

4. Transfer the parchment paper and loaf to the baking setup.

If baking with a cover: Cover and bake 25 minutes for the steam phase, then uncover and continue baking until the interior is 207°F (97°C) and the crust browns to your liking, usually 35 to 50 minutes total baking time.

If baking with a steam pan: Place the loaf in the oven, then pour 1½ cups (355 milliliters) of water into the steam pan. Bake 25 minutes without opening the oven for the steam phase, then continue baking until the interior is 207°F (97°C) and the crust browns to your liking, usually 35 to 50 minutes total baking time.

5. Cure in the cooling oven with the door ajar 5 to 10 minutes. Cool 1 to 2 hours on a rack.

Oatmeal Bread

Whenever I make this bread, it disappears shortly after it's cut. There is something in the combination of a custardy, open crumb and the creamy, sweet aroma of oats that draws people back for a second or third slice. The oats go stealth and disappear into the crumb, so this is a great loaf for pleasing a group with diverse preferences.

Although this recipe calls for a slow fermentation, we use a few rounds of folds at the beginning of the bulk fermentation. This assures the gluten is well developed to hold up the oats and to give us a light, open crumb.

MAKES ONE LOAF

TOTAL FLOUR: 450 GRAMS PLUS 150 GRAMS OATS (1 POUND PLUS 0.33 POUND)

PRE-FERMENTED FLOUR: 3%

HYDRATION: 83%

WHOLE GRAIN: 25%

DOUGH: 1119 GRAMS (2.5 POUNDS)

DAY 1

Mix the Dough (10 minutes)

1. Mix the flour in a large heavy bowl and set aside:

435 grams bread flour (3⅝ cups)

150 grams rolled oats (1½ cups)

9 grams salt (1½ teaspoons)

2. Whisk the wet ingredients together in a medium bowl until uniform:

495 grams water (2⅛ cups)

30 grams starter that floats (2 tablespoons)

3. Mix the wet ingredients into the flour, combining completely. The dry flour and oats need to be fully incorporated. The dough will seem very shaggy and wet, but do not add more flour. The oats will soak up a lot of water.

Bulk Fermentation (10 to 18 hours)

1. Cover the dough and let it rest at 60–70°F (16–21°C) for an hour. Then fold the dough with a wet spatula until it forms a ball and flip the ball. Rest another 15 to 30 minutes, then perform a set of stretch and folds. Repeat the rest and perform a final set of folds. The dough should now feel strong.

2. Continue resting undisturbed for 10 to 18 hours total until the dough is puffed up to two or three times its original size, is jiggly, and smells ready, with bubbles throughout. (If you plan on a loaf retard, end the bulk fermentation at two times its original size.)

continued

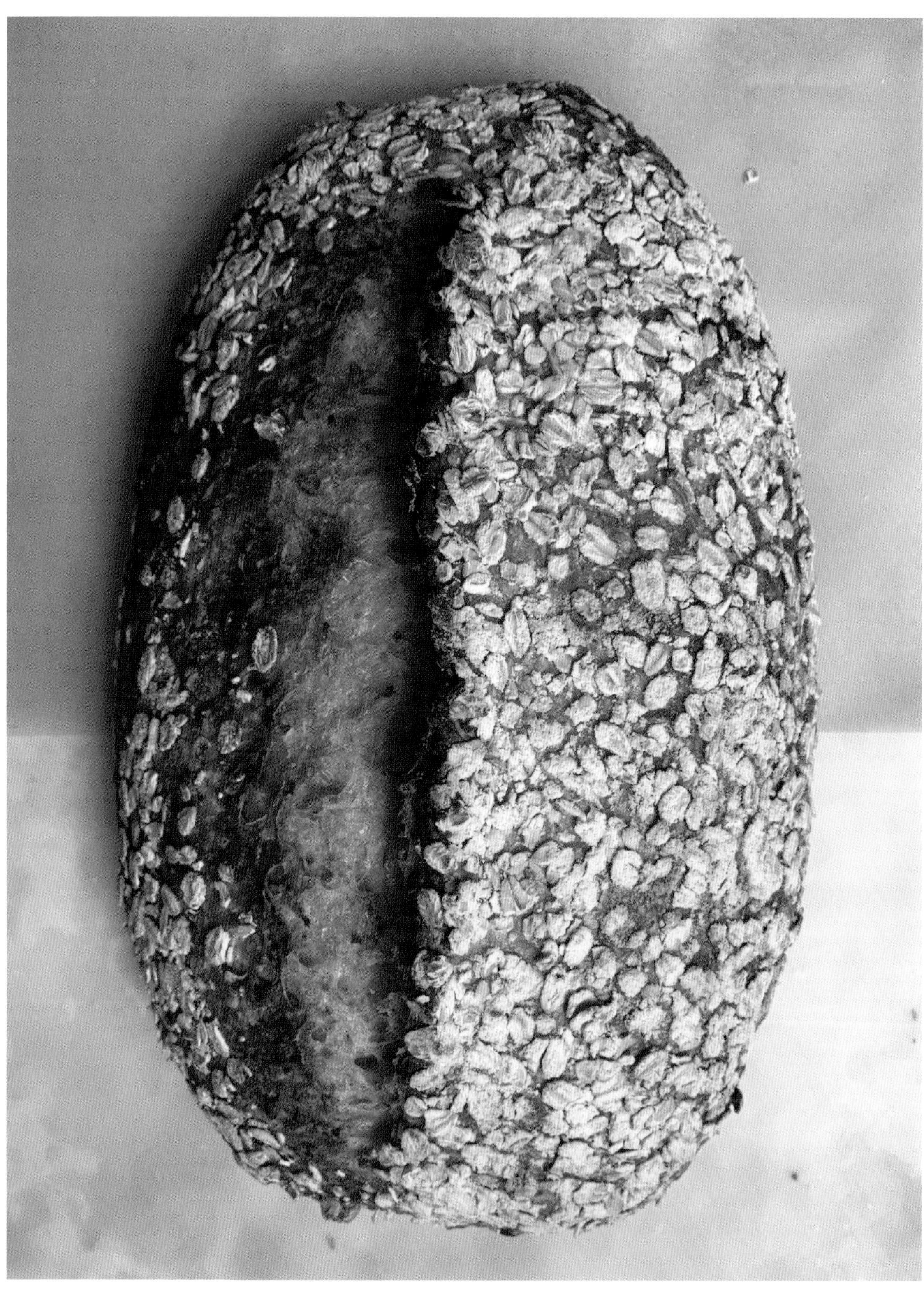

DAY 1 OR 2

Shape the Loaf (15 to 20 minutes)

1. Turn the dough out gently onto a damp counter. It should seem lively, wet, and bubbly.

2. Pre-shape: fold the dough into thirds, like a letter, fold in the other two sides, and flip the ball. Rotate it to tighten the skin. Let it rest it for 10 to 15 minutes.

3. Shape the dough into a ball or batard without deflating it and tighten the skin.

4. Prepare a wide bowl, plate, or platter sprinkled with rolled oats.

5. Brush off excess flour and mist the top and sides of the loaf lightly so the oats will stick.

6. Roll the loaf top side down onto the rolled oats until they adhere, then lift the loaf and place it in a well-floured proofing basket. Sprinkle rolled oats or flour around the edges where the dough tends to stick as it rises and lay parchment paper on top.

Proof the Loaf (30 minutes to 18 hours)

Cover and let it rest at room temperature for 30 to 90 minutes until the loaf seems risen and is jiggly like a water balloon. Or retard the loaf in the refrigerator for 4 to 18 hours.

DAY 2 OR 3

Bake the Loaf (1½ to 2 hours)

1. Preheat the oven to 475°F (246°C) 30 to 60 minutes before baking, and prepare your baking and steam setup.

2. Turn out the loaf and score it using kitchen shears to cut through the oats if needed.

3. Mist the loaf generously (optional) and transfer to the oven using your chosen baking and steam setup.

If baking with a cover: Cover and bake 25 minutes for the steam phase, then uncover and continue baking until the interior is 207°F (97°C) and the crust browns to your liking, usually 35 to 50 minutes total baking time.

If baking with a steam pan: Place the loaf in the oven, then pour 1½ cups (355 milliliters) of water into the steam pan. Bake 25 minutes without opening the oven for the steam phase, then continue baking until the interior is 207°F (97°C) and the crust browns to your liking, usually 35 to 50 minutes total baking time.

4. Cure in the cooling oven with the door ajar for 5 to 10 minutes. Cool 1 to 2 hours on a rack.

Pain de Campagne

A tender, fragrant, shiny crumb in a crackling, crisp crust makes this classic French boule a crowd-pleaser. It's the perfect base for French cheeses or pâtés or to dip in your cassoulet.

This delicious and beautiful bread is easy to make. For the most delectable flavor and crust, opt to retard the loaf as it proofs. The character of this bread really builds during the long fermentation, but its sourness is kept in check by keeping the temperature below 70°F (21°C) during the non-refrigerated fermentations to slow the growth of the lactic acid bacteria. To further avoid sourness, use a recently refreshed starter.

MAKES ONE LOAF

TOTAL FLOUR: 450 GRAMS (1 POUND)

PRE-FERMENTED FLOUR: 3%

HYDRATION: 82%

WHOLE GRAIN: 32%

DOUGH: 837 GRAMS (1.8 POUNDS)

DAY 1

Mix the Dough (10 minutes)

1. Mix the flour in a large heavy bowl and set aside:

300 grams bread flour (2½ cups)

45 grams whole rye flour (⅜ cup)

90 grams whole wheat flour (¾ cup)

9 grams diastatic barley malt flour (1 tablespoon)

9 grams salt (1½ teaspoons)

2. Whisk the wet ingredients together in a small bowl until uniform:

354 grams water (1½ cups)

30 grams starter that floats (2 tablespoons)

3. Mix the wet ingredients into the flour, combining completely. The dry flour needs to be fully incorporated. The dough will seem very shaggy and wet, but do not add more flour.

Bulk Fermentation (10 to 18 hours)

1. Cover the dough and let it rest at 60–70°F (16–21°C) for 30 to 60 minutes. Then fold the dough with a wet spatula until it forms a wet, loose ball and flip the ball.

2. Continue resting for 10 to 18 hours total until the dough is puffed up to two or three times its original size, is jiggly, and smells ready, with many large bubbles visible on the surface. (If you plan on retarding the loaf, end the bulk fermentation at two times its original size.)

continued

DAY 1 OR 2

Shape the Loaf (15 to 20 minutes)

1. Turn the dough out gently onto a damp counter. It should seem lively, strong, and bubbly.

2. Pre-shape: fold the dough into thirds, like a letter, fold in the other two sides, and flip the ball. Rotate it to tighten the skin. Let it rest for 10 to 15 minutes. It will seem like a big water balloon.

3. Shape the dough into a ball or batard without deflating it and tighten the skin.

4. Sprinkle the entire top and sides of the loaf with a generous coating of flour.

5. Place the loaf seam side up in a well-floured proofing basket. A linen cloth liner and generous flour help to keep this wet dough from sticking.

6. Sprinkle it with flour, especially around the edges. Lay parchment paper on top.

Proof the Loaf: (30 minutes to 18 hours)

Cover and let it rest at room temperature for 30 to 90 minutes until the loaf has risen and is jiggly like a water balloon. Since we extended the bulk fermentation and retained most of the loft in the dough, a short loaf proof is sufficient and will help us get good oven spring. Or retard the loaf in the refrigerator for 4 to 18 hours, provided you ended the bulk fermentation early.

DAY 2 OR 3

Bake the Loaf (1½ to 2 hours)

1. Preheat the oven to 475°F (246°C) 30 to 60 minutes before baking, and prepare your baking and steam setup.

2. Turn out the loaf and dust off excess flour. The loaf may spread out quite a lot.

3. Score the loaf and mist generously (optional).

4. Transfer the parchment paper and loaf to the baking setup.

If baking with a cover: Cover and bake 30 minutes for the steam phase, then uncover and continue baking until the interior is 207°F (97°C) and the crust browns to your liking, usually 35 to 50 minutes total baking time.

If baking with a steam pan: Place the loaf in the oven, then pour 1½ cups (355 milliliters) of water into the steam pan. Bake 30 minutes without opening the oven for the steam phase, then continue baking until the interior is 207°F (97°C) and the crust browns to your liking, usually 35 to 50 minutes total baking time.

5. Cure in the cooling oven with the door ajar 5 to 10 minutes. Cool 1 to 2 hours on a rack.

Walnut Bread

A showstopper with any cheese, this bread is particularly wonderful with blue cheese and some pears or apples. Whole wheat flour brings the extra depth of flavor to the crumb to complement the richness of the nuts.

Toast the walnuts for 10 minutes at 350°F (177°C) for even better flavor. This recipe is also spectacular with pecans.

MAKES ONE LOAF

TOTAL FLOUR: 450 GRAMS (1 POUND)

PRE-FERMENTED FLOUR: 3%

HYDRATION: 80%

WHOLE GRAIN: 33%

DOUGH: 972 GRAMS (2.1 POUNDS)

DAY 1

Mix the Dough (10 minutes)

1. Mix the flours in a large heavy bowl and set aside:

> **150 grams whole wheat flour (1¼ cups)**
>
> **285 grams bread flour (2⅜ cups)**
>
> **9 grams salt (1½ teaspoons)**

2. Whisk the wet ingredients together in a small bowl until uniform:

> **345 grams water (scant 1½ cups)**
>
> **30 grams starter that floats (2 tablespoons)**

3. Mix the wet ingredients into the flour, combining completely.

4. Fold in the nuts:

> **150 grams toasted walnuts or pecans (1½ cups)**

Bulk Fermentation (10 to 18 hours)

Cover the dough and let it rest at 60–70°F (16–21°C) for 10 to 18 hours until the dough is elastic and passes the windowpane test, the smell of flour is gone and it smells more like bread, and the dough jiggles in its bowl when you shake it. Pay attention near the end to prevent over-fermentation.

continued

DAY 1 OR 2

Shape the Loaf (20 minutes)

1. Turn the dough out gently onto a damp counter. It will seem wet and bubbly.
2. Pre-shape: fold the dough into thirds, like a letter, fold in the other two sides, and flip the ball. Rotate it to tighten the skin. Let it rest for 15 minutes.
3. Shape the dough into a ball or batard without deflating it, being gentle to the dough and poking back any nuts that try to escape.
4. Flip the loaf and slowly tighten the skin, being very careful not to tear the skin.
5. Flour the top generously and place the loaf seam side up in a well-floured proofing basket.
6. Sprinkle it with flour, especially around the edges where dough tends to stick as it rises. Lay parchment paper on top.

Proof the Loaf (1 to 18 hours)

Cover and let it rest at room temperature for 1 to 3 hours until the loaf has risen and is jiggly like a water balloon. Or retard the loaf in the refrigerator for 4 to 18 hours, provided you are confident the dough is not over-proofed.

DAY 2 OR 3

Bake the Loaf (1½ to 2 hours)

1. Preheat the oven to 475°F (246°C) 30 to 60 minutes before baking, and prepare your baking and steam setup.
2. Turn out the loaf and score it.
3. Mist the loaf generously and transfer to the oven using your chosen baking and steam setup.

If baking with a cover: Cover and bake 20 minutes for the steam phase, then uncover and continue baking until the interior is 207°F (97°C) and the crust browns to your liking, usually 35 to 50 minutes total baking time.

If baking with a steam pan: Place the loaf in the oven, then pour 1½ cups (355 milliliters) of water into the steam pan. Bake 20 minutes without opening the oven for the steam phase, then continue baking until the interior is 207°F (97°C) and the crust browns to your liking, usually 35 to 50 minutes total baking time.

4. Cure in the cooling oven with the door ajar for 5 to 10 minutes. Cool 1 to 2 hours on a rack.

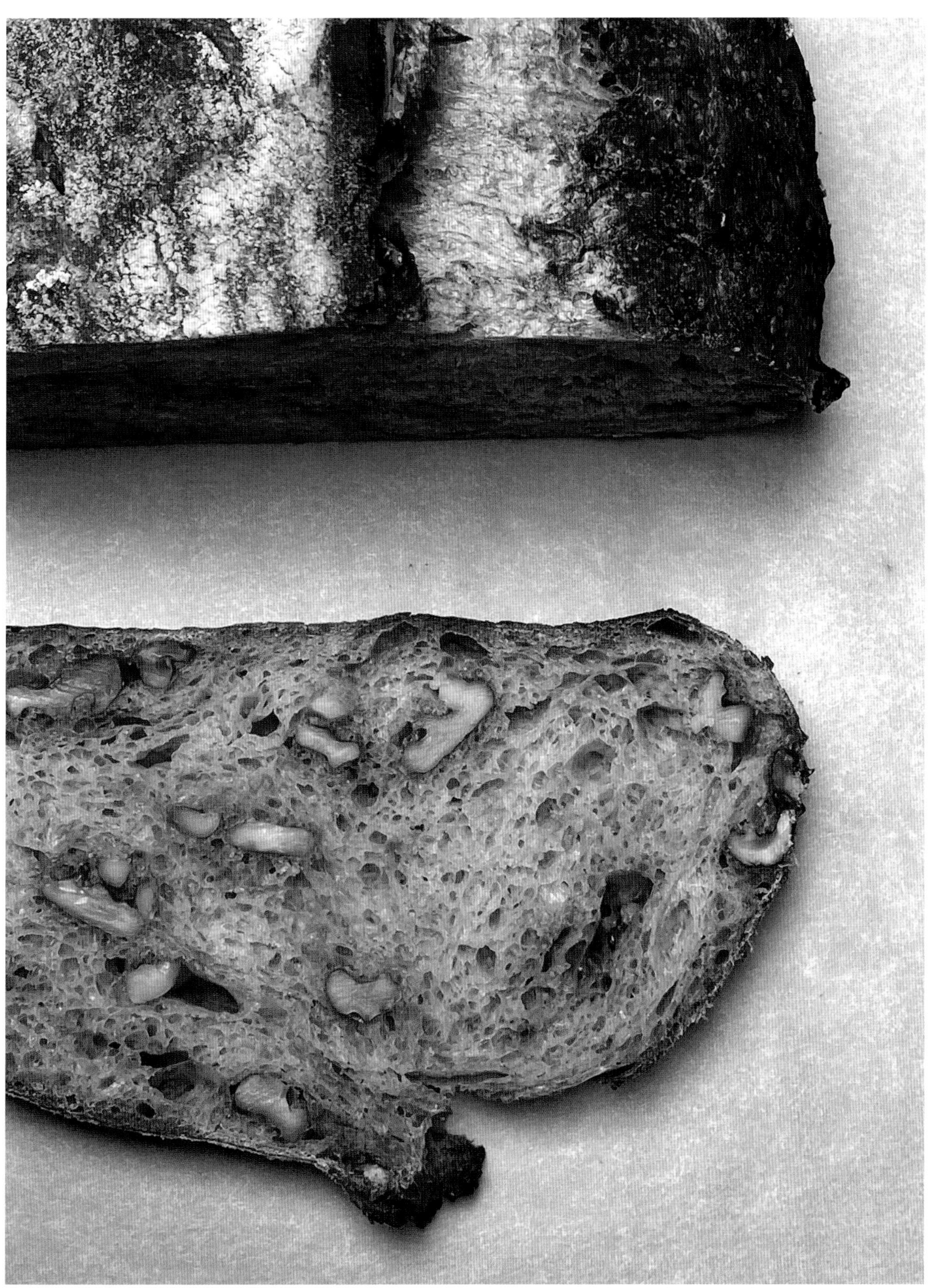

60% Whole Wheat

An easy and healthy everyday bread with great wheat and sourdough flavor thanks to the long fermentation. The open crumb is springy and chewy, and the crunchy crust carries intense flavor, partly due to the enzymes in the malted barley. Amazing for avocado toast, sandwiches, or dipping in fruity olive oil.

A gentle touch with the dough will keep the gluten network intact and will reward you with a light and open crumb. The bread pictured was made with a stone-ground white whole wheat, and its bran lacks the red pigment that gives much of the brown color to whole wheat bread.

MAKES ONE LOAF

FLOUR: 450 GRAMS (1 POUND)

PRE-FERMENTED FLOUR: 3%

HYDRATION: 85%

WHOLE GRAIN: 62%

DOUGH: 845 GRAMS (1.9 POUNDS)

DAY 1

Mix the Dough (10 minutes)

1. Mix the flours in a large heavy bowl and set aside:

269 grams whole wheat flour (2¼ cups)

157 grams bread flour (1¼ cups)

9 grams diastatic barley malt flour (1 tablespoon; optional)

9 grams salt (1½ teaspoons)

2. Whisk the wet ingredients together in a small bowl until uniform:

368 grams water (1½ cups plus 1 tablespoon)

30 grams starter that floats (2 tablespoons)

3. Mix the wet ingredients into the flour, combining completely.

Bulk Fermentation (10 to 18 hours)

1. Cover the dough and let it rest at 60–70°F (16–21°C) for an hour. Then use a wet spatula to fold the dough from the edges to the center all around the bowl, about 10 folds, until it forms a nice ball. Flip the ball.

2. Continue to let the dough rest undisturbed for a total of 10 to 18 hours until it is about double in volume, is elastic, the smell of flour is gone and it smells more like bread, and the dough jiggles in its bowl when you shake it. Pay attention near the end to prevent over-fermentation.

continued

DAY 1 OR 2

Shape the Loaf (20 minutes)

1. Turn the dough out gently onto a damp counter. It will seem wet and bubbly.
2. Pre-shape: fold the dough into thirds, like a letter, fold in the other two sides, and flip the ball. Rotate it to tighten the skin. Let it rest for 15 minutes.
3. Shape the dough into a ball or batard very gently without deflating it and tighten the skin, working slowly so as not to tear the skin.
4. Flour the top generously and place the loaf seam side up in a well-floured proofing basket. Sprinkle it with flour, especially around the edges. Lay parchment paper on top.

Proof the Loaf (1 to 12 hours)

Cover the loaf and let it rest at room temperature for 1 to 3 hours until it seems like a water balloon. Or retard the loaf in the refrigerator for 4 to 12 hours, provided you are confident the dough is not over-proofed.

DAY 2

Bake the Loaf (1½ to 2 hours)

1. Preheat the oven to 475°F (246°C) 30 to 60 minutes before baking, and prepare your baking and steam setup.
2. Turn out the loaf and score it.
3. Mist the loaf generously (optional) and transfer to the oven using your chosen baking and steam setup.

If baking with a cover: Cover and bake 20 minutes for the steam phase, then uncover and continue baking until the interior is 207°F (97°C) and the crust browns to your liking, usually 35 to 50 minutes total baking time.

If baking with a steam pan: Place the loaf in the oven, then pour 1½ cups (355 milliliters) of water into the steam pan. Bake 20 minutes without opening the oven for the steam phase, then continue baking until the interior is 207°F (97°C) and the crust browns to your liking, usually 35 to 50 minutes total baking time.

4. Cure in the cooling oven with the door ajar for 5 to 10 minutes. Cool 1 to 2 hours on a rack.

Sesame Grano Duro

Durum is a lovely flour to work with, and it forms a satiny, soft dough and crumb. Its amber color derives from the carotenoids that give durum its antioxidant-rich nutritional punch. This grain, a high-protein descendant of emmer, imparts a lovely, creamy flavor and velvety texture to the bread.

If you are unable to source whole grain durum flour you can substitute a refined durum flour such as Italian "semola rimacinata di grano duro" flour. In that case, the dough will handle like a white flour dough, but with less strength, and you could use 25 to 45 grams less water. Avoid using semolina, which is too coarsely ground for this recipe. Durum, although very high in protein, makes a delicate gluten network that resents kneading.

MAKES ONE LOAF

TOTAL FLOUR: 450 GRAMS (1 POUND)
PRE-FERMENTED FLOUR: 3%
HYDRATION: 80%
WHOLE GRAIN: 50%
DOUGH: 821 GRAMS (1.8 POUNDS)

DAY 1

Mix the Dough (10 minutes)

1. Mix the flours in a large heavy bowl and set aside:

217 grams whole grain durum flour (1¾ cups plus 1 tablespoon)

210 grams bread flour (1¾ cups)

9 grams diastatic barley malt flour (1 tablespoon)

9 grams salt (1½ teaspoons)

2. Whisk the wet ingredients together in a small bowl until uniform:

346 grams water (scant 1½ cups)

30 grams starter that floats (2 tablespoons)

3. Mix the wet ingredients into the flour, combining completely. The dough should seem very dry and tough, but all the dry flour needs to be incorporated. If necessary, rest the dough for 10 minutes and then finish mixing.

Bulk Fermentation (10 to 18 hours)

Cover the dough and let it rest at 60–70°F (16–21°C) for 10 to 18 hours until the dough is puffed up, jiggly, and smells like bread or a bakery.

continued

DAY 1 OR 2

Shape the Loaf (20 minutes)

1. Turn the dough out gently onto a damp counter. It should seem very wet and bubbly.

2. Pre-shape: fold the dough into thirds, like a letter, fold in the other two sides, and flip the ball. Rotate it to tighten the skin. Let it rest for 15 minutes.

3. Lay out a square of waxed or parchment paper and sprinkle with:

> **10 to 40 grams sesame seeds (1 to 4 tablespoons)**

4. Shape the dough into a ball or batard very gently without deflating it, and do not expect a lot of stretch from the dough.

5. Flip the loaf and tighten the skin, being very gentle so as not to tear the skin.

6. Gently rub the entire top and sides of the loaf with a light coating of oil so it won't stick to the waxed paper during proofing.

7. Place the loaf top side down onto the sesame waxed paper.

8. Transfer the paper and dough to the proofing basket, keeping the paper under the dough.

9. Sprinkle seeds or flour around the edges where dough tends to stick as it rises.

10. Sprinkle flour on top and lay parchment paper over the dough.

Proof the Loaf (1 to 12 hours)

Cover and let it rest at room temperature for 1 to 3 hours until the loaf seems like a water balloon. Or retard the loaf in the refrigerator for 4 to 12 hours, provided you are confident the dough is not over-proofed.

DAY 2

Bake the Loaf (1½ to 2 hours)

1. Preheat the oven to 475°F (246°C) 30 to 60 minutes before baking, and prepare your baking and steam setup.

2. Turn out the loaf and remove the waxed paper.

3. Score the loaf and mist generously.

4. Transfer the parchment paper and loaf to the baking setup.

If baking with a cover: Cover and bake 20 minutes for the steam phase, then uncover and continue baking until the interior is 207°F (97°C) and the crust browns to your liking, usually 35 to 50 minutes total baking time.

If baking with a steam pan: Place the loaf in the oven, then pour 1½ cups (355 milliliters) of water into the steam pan. Bake 20 minutes without opening the oven for the steam phase, then continue baking until the interior is 207°F (97°C) and the crust browns to your liking, usually 35 to 50 minutes total baking time.

5. Watch that the seeds are not overbrowning toward the end. Tent with foil to protect them if necessary.

6. Cure in the cooling oven with the door ajar for 5 to 10 minutes. Cool 1 to 2 hours on a rack.

Multigrain Bread

This is pure sustenance when you feel overstretched or energy depleted and want something to fortify your soul. It takes a little more work to measure out, but the rewards are great. Excellent as the accompaniment to a hearty soup or salad, or with a piece of sharp aged Gouda or cheddar. The oats and corn disappear into the crumb but lend their sweet aromas. The millet contributes a festive, confetti look and a delightful crunch to the finished bread.

The dough is quite easy to mix and handle, and it requires very little hands-on time. We bake at a lower temperature to prevent an overly brown crust.

MAKES ONE LOAF

FLOUR: 450 GRAMS (1 POUND)

PRE-FERMENTED FLOUR: 3%

HYDRATION: 72%

WHOLE GRAIN: 67%

DOUGH: 823 GRAMS (1.8 POUNDS)

DAY 1

Mix the Dough (15 minutes)

1. Mix the flours in a large heavy bowl and set aside:

135 grams bread flour (1⅛ cups)

120 grams whole wheat flour (1 cup)

60 grams corn flour (½ cup)

60 grams rolled oats (⅔ cup)

60 grams millet (⅓ cup)

9 grams salt (1½ teaspoons)

2. Whisk the wet ingredients together in a small bowl until uniform:

309 grams water (1¼ cups plus 1 tablespoon)

40 grams honey (2 tablespoons)

30 grams starter that floats (2 tablespoons)

3. Mix the wet ingredients into the flour, combining completely. The dough should seem like a wet and lumpy bowl of oatmeal. Cover the dough and let it rest for 30 to 60 minutes.

Bulk Fermentation (10 to 18 hours)

1. Perform spatula folds until the dough comes together in a loose ball. It will seem soft, but do not add more flour because the oats and millet will soak up some water.

continued

2. Continue to rest the dough at 60–70°F (16–21°C) for 10 to 18 hours until it is doubled in volume with many bubbles and begins to be jiggly. Expect only a small windowpane. Pay attention near the end to prevent over-fermentation and to end the fermentation early if you plan to retard the loaf.

DAY 1 OR 2

Shape the Loaf (20 minutes)

1. Handle the dough minimally and gently during shaping to achieve a light, lofty crumb. Turn the dough out onto a lightly misted counter and pre-shape: fold the dough into thirds, like a letter, fold in the other two sides, and flip the ball. Rotate it to tighten the skin. Let it rest 15 minutes until it relaxes.

2. Shape into a boule or batard and gently tighten the skin, watching closely for tearing. Flour the top generously.

3. Place seam side up in a well-floured proofing basket. Sprinkle it with flour, especially around the edges. Lay parchment paper on top.

Proof the Loaf (1 to 12 hours)

Cover and let it rest at room temperature until the loaf has risen and is jiggly like a water balloon (1 to 3 hours). Or retard in the refrigerator for 4 to 12 hours, provided you are confident the dough is not over-proofed.

DAY 2

Bake the Loaf (1½ to 2 hours)

1. Preheat the oven to 450°F (232°C) 30 to 60 minutes before baking, and prepare your baking and steam setup.

2. Turn out the loaf and score it.

3. Mist the loaf generously and transfer to the oven using your chosen baking and steam setup.

If baking with a cover: Cover and bake 25 minutes for the steam phase, then uncover and continue baking until the interior is 207°F (97°C) and the crust browns to your liking, usually 35 to 50 minutes total baking time.

If baking with a steam pan: Place the loaf in the oven, then pour 1½ cups (355 milliliters) of water into the steam pan. Bake 25 minutes without opening the oven for the steam phase, then continue baking until the interior is 207°F (97°C) and the crust browns to your liking, usually 35 to 50 minutes total baking time.

4. Cure in the cooling oven with the door ajar for 5 to 10 minutes. Cool 1 to 2 hours on a rack.

Beach Rolls with Rosemary and Gruyère

Fragrant rosemary and sharp Gruyère are rolled inside these mini batard rolls for a hearty snack that's easy to eat at the beach (or on the way there!). Stuff with prosciutto and roasted peppers for an easy lunch. You can mix the dough in the morning and shape the rolls in the evening, then retard overnight to bake right when you wake up on beach day. By the time you've packed the boogie boards and sunscreen, they'll be ready to pop in the picnic basket. I love rolls for the beach because their individual crusts keep them from getting stale in the fresh ocean breeze.

The rye makes the dough a bit sticky and quick to ferment, but the flavor it adds can't be beat. There's plenty of bread flour to give the rolls a nice loft.

MAKES EIGHT ROLLS

TOTAL FLOUR: 450 GRAMS (1 POUND)

PRE-FERMENTED FLOUR: 3%

HYDRATION: 78%

WHOLE GRAIN: 50%

DOUGH: 986 GRAMS (2.2 POUNDS)

DAY 1

Mix the Dough (10 minutes)

1. Mix the flours in a large heavy bowl and set aside:

> **210 grams bread flour (1¾ cups)**
>
> **126 grams whole wheat flour (1 cup plus 1 tablespoon)**
>
> **90 grams whole grain rye flour (¾ cup)**
>
> **9 grams diastatic barley malt flour (1 tablespoon)**
>
> **9 grams salt (1½ teaspoons)**

2. Whisk the wet ingredients together in a small bowl until uniform:

> **336 grams water (1⅜ cups)**
>
> **30 grams starter that floats (2 tablespoons)**

3. Mix the wet ingredients into the flour, combining completely.

Bulk Fermentation (8 to 12 hours)

1. Cover the dough and let it rest at 60–70°F (16–21°C).

2. After 1 hour, do a set of stretch and folds (optional, but makes shaping easier).

3. Rest the dough for a total of 8 to 12 hours until it is elastic and passes the windowpane test, the smell of flour is gone and it smells more like bread, and the dough just starts being able to jiggle in its bowl when you shake it. Pay attention near the end to prevent over-fermentation, which would make it very hard to shape the rolls.

continued

DAY 1 OR 2

Shape the Rolls (20 to 30 minutes)

1. Mix together in a small bowl:

   170 grams grated Gruyère cheese (6 ounces), or substitute any sharp hard cheese

   6 grams chopped fresh rosemary leaves (2 tablespoons) from 4 to 6 sprigs

   freshly ground black pepper, to taste

2. Turn the dough out onto a floured counter and stretch into a rough rectangle shape, about 8 by 10 inches (20 by 25 centimeters). It will feel very soft, sticky, and wet.

3. Sprinkle the cheese mixture evenly onto the dough and press it in to adhere.

4. Cut the rectangle into eight equal pieces with a bench knife.

5. Pre-shape each piece: fold the dough into thirds, like a letter, fold in the other two sides, and flip the ball, tucking all the filling inside the ball. Rotate it to tighten the skin. Let it rest, seam side down, for 10 to 15 minutes. You will need to use some flour, and you may find it helpful to use the bench knife to help lift the sides of the dough to fold it over the center. Let it rest 5 to 10 minutes.

6. Gently form each ball into a mini-batard by simply tightening the skin with your bench knife on two sides.

7. Arrange the rolls on parchment paper on a baking sheet with room to expand, sprinkle with flour, and lay parchment paper on top. Cover them lightly with a damp tea towel or a plastic bag to keep them from drying while proofing.

Proof the Rolls (30 minutes to 12 hours)

Let the rolls rest at 65–70°F (16–21°C) for 30 to 90 minutes until slightly risen. Or retard the rolls in the refrigerator for 3 to 12 hours.

DAY 2

Bake the Rolls (1½ hours)

1. Preheat the oven to 400°F (204°C) for 30 to 60 minutes before you want to bake, and prepare your baking and steam setup.

2. When the oven is ready, score the rolls and mist them heavily with water (optional), then place the pan in the oven (or slide them, parchment paper and all, onto a hot baking stone).

If baking with a cover: Cover and bake 10 minutes for the steam phase, then uncover and continue baking until the interior is 207°F (97°C) and the crust browns to your liking, usually 25 to 40 minutes total baking time.

If baking with a steam pan: Place the loaf in the oven, then pour 1 cup (237 milliliters) of water into the steam pan. Bake 10 minutes without opening the oven for the steam phase, then continue baking until the interior is 207°F (97°C) and the crust browns to your liking, usually 25 to 40 minutes total baking time.

3. Cool on a rack for at least 15 minutes before serving.

Spelt-o-licious

This loaf showcases the sweet aroma and nutty flavor of fresh whole grain spelt flour in a beautiful artisan boule with a crunchy crust. A little bit of bread flour and a high hydration level makes for an airy crumb. Perfect slathered with honey and nut butter, or toasted with avocado and melted Monterey Jack cheese.

The dough will seem all wrong when you first mix it, but it will transform itself by the time you shape the loaf. You can switch out the spelt for another weak-gluten ancient grain if you like, such as einkorn, emmer, or khorasan.

MAKES ONE LOAF

TOTAL FLOUR: 450 GRAMS (1 POUND)

PRE-FERMENTED FLOUR: 3%

HYDRATION: 75%

WHOLE GRAIN: 70%

DOUGH: 837 GRAMS (1.8 POUNDS)

DAY 1

Mix the Dough (10 minutes)

1. Mix the flours in a large heavy bowl and set aside:

120 grams bread flour (1 cup)

315 grams whole spelt flour (2⅝ cups)

9 grams salt (1½ teaspoons)

2. Whisk the wet ingredients together in a small bowl until uniform:

323 grams water (1⅜ cups)

40 grams honey (2 tablespoons)

30 grams starter that floats (2 tablespoons)

3. Mix the wet ingredients into the flour, combining completely. It will mix easily and look like a bowl of oatmeal, but don't add extra flour.

Bulk Fermentation (8 to 12 hours)

1. Cover the dough and let it rest at 60–70°F (16–21°C) for an hour. Then use a wet spatula to fold the dough from the edges to the center all around the bowl, about 20 folds, until it forms a nice ball. Flip the ball.

2. Continue to rest the dough for a total of 8 to 12 hours until it is bubbly, has risen two or three times the original volume, and is jiggly. It should be able to make a decent windowpane and smell like grassy bread. Pay attention near the end to prevent over-fermentation.

DAY 1 OR 2

Shape the Loaf (20 minutes)

1. Turn the dough out gently onto a damp counter. It will seem wet, rough, and bubbly.

2. Pre-shape: fold the dough into thirds, like a letter, fold in the other two sides, and flip the ball. Rotate it to tighten the skin. Let it rest for 15 minutes.

3. Shape the dough into a ball or batard very gently without deflating it, and tighten the skin.

4. Flour the top generously and place the loaf seam side up in a well-floured proofing basket. Sprinkle it with flour, especially around the edges. Lay parchment paper on top.

Proof the Loaf (1 to 12 hours)

Cover and let it rest at 65–70°F (18–21°C) for 1 to 3 hours until the loaf seems jiggly like a water balloon. Or retard the loaf in the refrigerator for 4 to 12 hours, provided you are confident the dough is not over-proofed.

DAY 2

Bake the Loaf (1½ to 2 hours)

1. Preheat the oven to 475°F (246°C) 30 to 60 minutes before baking, and prepare your baking and steam setup.

2. Turn out the loaf and score it shallowly. Don't expect a lot of oven spring.

3. Mist the loaf (optional) and transfer to the oven using your chosen baking and steam setup.

If baking with a cover: Cover and bake 25 minutes for the steam phase, then uncover and continue baking until the interior is 207°F (97°C) and the crust browns to your liking, usually 35 to 50 minutes total baking time.

If baking with a steam pan: Place the loaf in the oven, then pour 1½ cups (355 milliliters) of water into the steam pan. Bake 25 minutes without opening the oven for the steam phase, then continue baking until the interior is 207°F (97°C) and the crust browns to your liking, usually 35 to 50 minutes total baking time.

4. Cure in the cooling oven with the door ajar for 5 to 10 minutes. Cool 1 to 2 hours on a rack.

Whole Grain Uncut

Whether you are trying to avoid refined flour or you just enjoy the vibrant, complex flavors imparted by whole grains, this bread is sure to please. A caramelized, crisp but tender crust encases a soft, springy interior with a considerably open crumb for bread with no refined flour. Honey gives a suggestion of sweetness and keeps the bread soft and moist.

This sticky, delicate dough demands to be handled gently and stretched slowly, with wet hands and a bench knife or spatula. The recipe involves very little handling so that the large amount of natural bran in the dough does not cut the gluten network. The result is a nice, light crumb and a great rise.

MAKES ONE LOAF

TOTAL FLOUR: 450 GRAMS (1 POUND)

PRE-FERMENTED FLOUR: 3%

HYDRATION: 85%

WHOLE GRAIN: 97%

DOUGH: 882 GRAMS (1.9 POUNDS)

DAY 1

Mix the Dough (10 minutes)

1. Mix the flours in a large heavy bowl and set aside:

90 grams whole rye flour (¾ cup)

345 grams whole wheat flour (2⅞ cups)

9 grams salt (1½ teaspoons)

2. Whisk the wet ingredients together in a small bowl until uniform:

368 grams water (1½ cups plus 1 tablespoon)

40 grams honey (2 tablespoons)

30 grams starter that floats (2 tablespoons)

3. Mix the wet ingredients into the flour, combining completely.

Bulk Fermentation (10 to 18 hours)

Cover the dough and let it rest at 60–70°F (16–21°C) for 10 to 18 hours until it is doubled in volume with many bubbles and begins to be jiggly. Expect only a small windowpane. Pay attention near the end to prevent over-fermentation.

DAY 1 OR 2

Shape the Loaf (20 minutes)

1. Handle the dough minimally and gently during shaping to achieve a light, lofty crumb. Turn the dough out onto a lightly misted counter and pre-shape: fold the dough into thirds, like a letter, fold in the other two sides, and flip the ball. Rotate it to tighten the skin. Let it rest 15 minutes until it relaxes.

2. Shape into a boule or batard and gently tighten the skin, watching closely for tearing. Flour the top generously.

3. Place the loaf seam side up in a well-floured proofing basket. Sprinkle it with flour, especially around the edges. Lay parchment paper on top.

Proof the Loaf (1 to 12 hours)

Cover and let it rest at room temperature until risen and jiggly like a water balloon (1 to 3 hours). Or retard the loaf in the refrigerator for 4 to 12 hours, provided you are confident the dough is not over-proofed.

DAY 2

Bake the Loaf (1½ to 2 hours)

1. Preheat the oven to 450°F (232°C) 30 to 60 minutes before baking, and prepare your baking and steam setup.

2. Turn out the loaf and score it.

3. Mist the loaf generously and transfer to the oven using your chosen baking and steam setup.

If baking with a cover: Cover and bake 20 minutes for the steam phase, then uncover and continue baking until the interior is 207°F (97°C) and the crust browns to your liking, usually 35 to 50 minutes total baking time.

If baking with a steam pan: Place the loaf in the oven, then pour 1½ cups (355 milliliters) of water into the steam pan. Bake 20 minutes without opening the oven for the steam phase, then continue baking until the interior is 207°F (97°C) and the crust browns to your liking, usually 35 to 50 minutes total baking time.

4. Cure in the cooling oven with the door ajar for 5 to 10 minutes. Cool 1 to 2 hours on a rack.

Rye Raisin

This rye will take you somewhere deep—there's something very satisfying and comforting about the combination of rye and raisins. Pair it with some fresh cream cheese or brie, salty corned beef, or just eat it with butter.

It's among the easiest recipes to make because after stirring up the dough there's nothing at all to do until it's time to shape the loaf. The trick to this light and lofty rye is to leave it alone completely during bulk fermentation. We bake it at a lower temperature with a longer steam phase to slow down the browning of the crust. For a very special bread, substitute 1 cup (190 grams) of fresh blueberries in place of the raisins.

MAKES ONE LOAF

TOTAL FLOUR: 450 GRAMS (1 POUND)

PRE-FERMENTED FLOUR: 3%

HYDRATION: 80%

WHOLE GRAIN: 63%

DOUGH: 979 GRAMS (2.2 POUNDS)

DAY 1

Mix the Dough (10 minutes)

1. Mix the flours in a large heavy bowl and set aside:

135 grams whole rye flour (1⅛ cups)

150 grams whole wheat flour (1¼ cups)

150 grams bread flour (1¼ cups)

9 grams salt (1½ teaspoons)

2. Whisk the wet ingredients together in a small bowl until uniform:

345 grams water (scant 1½ cups)

40 grams honey (2 tablespoons)

30 grams starter that floats (2 tablespoons)

3. Mix the wet ingredients into the flour, combining completely. Let it sit 10 minutes, then mix again with about 10 spatula folds until it's uniform.

Bulk Fermentation (10 to 18 hours)

Cover the dough and let it rest at 60–70°F (16–21°C) for 10 to 18 hours until it is doubled in volume with many bubbles and begins to be jiggly. Expect only a small windowpane. Pay attention near the end to prevent over-fermentation.

continued

Shape the Loaf (20 minutes)

1. Handle the dough minimally and gently during shaping to achieve a light, lofty crumb. Turn the dough out onto a lightly misted counter and stretch into a rough rectangle about 8 by 10 inches (20 by 25 centimeters).

2. Add:

¾ cup raisins or dried currants (120 grams)

3. Sprinkle the raisins or dried currants over the dough, then pre-shape. Fold the dough into thirds, like a letter, fold in the other two sides, and flip the ball. Rotate it to tighten the skin. Let it rest seam side down 15 minutes until it relaxes.

4. Shape into a boule or batard and gently tighten the skin, watching closely for tearing. Flour the top generously.

5. Place seam side up in a well-floured proofing basket. Sprinkle it with flour, especially around the edges. Lay parchment paper on top.

Proof the Loaf (1 to 12 hours)

Cover and let it rest at room temperature until risen and jiggly like a water balloon (1 to 3 hours). Or retard the loaf in the refrigerator for 4 to 12 hours, provided you are confident the dough is not over-proofed.

Bake the Loaf (1½ to 2 hours)

1. Preheat the oven to 450°F (232°C) 30 to 60 minutes before baking, and prepare your baking and steam setup.

2. Turn out the loaf and score it.

3. Mist the loaf generously if desired and transfer to the oven using your chosen baking and steam setup.

If baking with a cover: Cover and bake 30 minutes for the steam phase, then uncover and continue baking until the interior is 207°F (97°C) and the crust browns to your liking, usually 35 to 50 minutes total baking time.

If baking with a steam pan: Place the loaf in the oven, then pour 1½ cups (355 milliliters) of water into the steam pan. Bake 30 minutes without opening the oven for the steam phase, then continue baking until the interior is 207°F (97°C) and the crust browns to your liking, usually 35 to 50 minutes total baking time.

4. Cure in the cooling oven with the door ajar for 5 to 10 minutes. Cool 1 to 2 hours on a rack.

The Bomb

A really excellent bread for tearing by hand and passing around the table or picnic blanket. A very hot initial temperature mimics a wood-fired oven, making a crunchy, caramelized crust. It's a perfect foil to the open structured, creamy, and fragrant crumb. For the very best flavor, make sure to leave some time for the cold retard during the bulk fermentation.

The free-form batards require very little handling, but this is not a beginner bread. When you see the loose, wet dough you'll be glad not to have to handle it much. You will be rewarded for preserving all those big bubbles when you break open the loaf. This recipe is best baked on a stone or sheet, and with a steam pan.

MAKES TWO SMALL LOAVES

TOTAL FLOUR: 450 GRAMS (1 POUND)
PRE-FERMENTED FLOUR: 3%
HYDRATION: 85%
WHOLE GRAIN: 25%
DOUGH: 842 GRAMS (1.9 POUNDS)

DAY 1

Mix the Dough (10 minutes)

1. Measure the flour into a large heavy bowl and set aside:

23 grams rye (3 tablespoons)

90 grams whole wheat flour (¾ cup)

322 grams bread flour (2¾ cups)

9 grams salt (1½ teaspoons)

2. Whisk the wet ingredients together in a small bowl until uniform:

368 grams water (1½ cups plus 1 tablespoon)

30 grams starter that floats (2 tablespoons)

3. Mix the wet ingredients into the flour, combining completely. The dry flour needs to be fully incorporated. The dough will seem very soft and wet, but do not add more flour.

continued

Bulk Fermentation (12 to 24 hours)

1. Cover the dough and let it rest at 60–70°F (16–21°C) for 30 to 60 minutes. Then fold the dough with a wet spatula until it forms a smooth, loose ball and flip the ball.

2. Rest the dough for 12 to 24 hours total, starting at 60–70°F (16–21°C) for 4 to 12 hours until it is very puffed up to 2 or 2½ times its original size, is very jiggly and smells ready, and with many large bubbles visible on the surface.

3. Then refrigerate the dough for 4 to 12 hours.

DAY 2

Shape the Loaves (10 minutes)

1. Prepare a parchment paper–lined peel or baking sheet for resting the loaves.

2. Turn the dough out onto a floured counter. It should seem lively, fragrant, and very wet and bubbly. Handle minimally to achieve an open crumb and use plenty of flour on the outside of the loaves while you are shaping, but dust off any that will end up in the interior of the loaf.

3. Divide the dough in two with a floured bench knife and pull the edges a bit as needed to make each a symmetrical shape. Fold each in thirds, like a letter, and sprinkle them generously with flour. Roll them over so the seam is on the bottom. Rest a couple of minutes.

4. Sprinkle flour generously on top of the loaves and tighten with a bench knife.

5. Transfer them seam side down onto the parchment paper—lined peel or baking sheet with plenty of space between them for rising and spreading. Gently stretch them to 12 inches (30 centimeter) long, aiming for an even width throughout the loaf, and dust heavily with flour.

Proof the Loaves (1 to 2 hours)

1. Cover with a tented towel, roasting pan, or puffed-up plastic bag to prevent drying. The dough will stick mercilessly to anything touching it.

2. Rest the dough at room temperature for 1 to 2 hours until the loaves are risen and are jiggly like a water balloon.

Bake the Loaves (1½ to 2 hours)

1. Preheat the oven to 525°F (274°C) 60 minutes before baking, and prepare your baking and steam setup.

2. Cut the parchment paper between the loaves if desired. Transfer the parchment paper and loaves to the oven and, wearing a hot mitt, pour 1½ cups (355 milliliters) of water into the steam pan and quickly close the oven.

3. Bake 15 minutes with steam, then remove the steam pan if it still has water in it and turn the oven down to 475°F (246°C). Bake until 207°F (97°C) and the crust browns to your liking (a deep cocoa brown is divine), usually 20 to 30 minutes total time. Check every 5 minutes after the steam phase because the crust will color very quickly once it begins to brown.

4. Cure in the cooling oven with the door ajar 5 to 10 minutes. Cool 1 to 2 hours on a rack.

Olive Bats

Kalamata olives nestle in a soft, open crumb that's encased in a dark, crunchy crust. Don't expect these loaves to be around long after baking them. They are amazing enjoyed with a spit-roasted chicken and a tomato-cucumber salad, a combination that makes for an easy dinner or picnic lunch.

The dough is shaped into two free-form "bats" that are best baked on a stone or sheet, and with a steam pan. To achieve a rustic crust, we will mimic a wood-fired oven by placing the loaf in a blisteringly hot oven, then turning down the temperature after the first few minutes.

MAKES TWO SMALL LOAVES

TOTAL FLOUR: 450 GRAMS (1 POUND)

PRE-FERMENTED FLOUR: 3%

HYDRATION: 78%

WHOLE GRAIN: 15%

DOUGH: 980 GRAMS (2.2 POUNDS)

DAY 1

Mix the Dough (10 minutes)

1. Measure the flour into a large heavy bowl and set aside:

   68 grams whole wheat flour (½ cup plus 1 tablespoon)

   367 grams bread flour (3 cups plus 1 tablespoon)

   9 grams salt (1½ teaspoons)

2. Whisk the wet ingredients together in a small bowl until uniform:

   336 grams water (scant 1½ cups)

   30 grams starter that floats (2 tablespoons)

3. Mix the wet ingredients into the flour, combining completely. The dry flour needs to be fully incorporated. The dough will seem very shaggy and wet, but do not add more flour.

Bulk Fermentation (10 to 18 hours)

1. Cover the dough and let it rest at 60–70°F (16–21°C) for 30 to 60 minutes.

2. Add:

> **170 grams Kalamata olives, well drained and pitted (6 ounces)**

3. Fold the dough with a wet spatula or hands to incorporate the olives—they will try to escape! Form a loose ball and flip the ball. Don't worry if the olives aren't well distributed because the next step will mix them in more.

4. Cover and let it rest for 30 to 60 minutes, then perform a set of stretch and folds. Repeat if the dough relaxed right away. Continue resting undisturbed for 10 to 18 hours total until the dough is very puffed up to three times its original size, is very jiggly and smells ready, and has many large bubbles visible on the surface. The dough could be refrigerated at this point for up to 12 hours.

DAY 2

Shape the Loaves (5 minutes)

1. Prepare a parchment paper–lined peel or baking sheet for resting the loaves.

2. Turn the dough out gently onto a damp counter. It should seem lively, wet, and bubbly.

3. Divide the dough in two with a damp bench knife and pull the edges a bit as needed to make each a symmetrical shape. Fold each in thirds, like a letter.

4. Sprinkle them generously with flour and transfer them seam side down onto the parchment paper–lined peel or baking sheet with space between them for rising.

5. Sprinkle flour generously on top of the loaves and gently stretch them to 12 inches (30 centimeters) long.

Proof the Loaves (1 to 3 hours)

1. Lay parchment paper over the loaves and cover with a towel, roasting pan, or plastic bag to prevent drying.

2. Let them rest at room temperature for 1 to 3 hours until the loaves have risen and are jiggly like a water balloon.

Bake the Loaves (1½ hours)

1. Preheat the oven to 500°F (260°C) 45 to 60 minutes before baking, and prepare your baking and steam setup.

2. Transfer the parchment paper and loaf to the oven and, wearing a hot mitt, pour 1 cup (237 milliliters) of water into the steam pan and quickly close the oven.

3. Bake 15 minutes with steam, then remove the steam pan if it still has water in it and turn the oven down to 475°F (246°C). Bake until the interior is 207°F (97°C) and the crust browns to your liking (a deep cocoa brown is divine), usually 30 to 45 minutes total time.

4. Cure in the cooling oven with the door ajar 5 to 10 minutes. Cool 1 to 2 hours on a rack.

Lazy Bones Bread

Try this recipe when you simply cannot be bothered, but you are jonesing for the smell of bread baking, the sound of the crust crackling as it cools, and the unparalleled flavor of freshly baked bread slathered with butter. The recipe calls for 15 percent whole grain flour of your choice. I like to use white whole wheat, but you can try spelt, durum, einkorn, emmer, rye, corn, oat, quinoa, buckwheat, barley, amaranth—anything goes. It will be delicious.

Pay attention to the instructions because some things are a little different as we pare down the hands-on time to the barest minimum. If you want to eat it the same day, start early in the morning and ferment it closer to 70°F (21°C).

MAKES ONE LOAF

TOTAL FLOUR: 450 GRAMS (1 POUND)
PRE-FERMENTED FLOUR: 3%
HYDRATION: 80%
WHOLE GRAIN: 15%
DOUGH: 819 GRAMS (1.8 POUNDS)

DAY 1

Mix the Dough (10 to 20 minutes)

1. Mix the flours in a large heavy bowl and set aside:

> **68 grams whole grain flour of your choice (½ cup plus 1 tablespoon)**
>
> **367 grams all-purpose flour (3 cups plus 1 tablespoon)**
>
> **9 grams salt (1½ teaspoons)**

2. Whisk the wet ingredients together in a small bowl until uniform:

> **345 grams water (scant 1½ cups)**
>
> **30 grams starter that floats (2 tablespoons)**

3. Mix the wet ingredients into the flour, combining completely, and if you have time, set aside for 10 minutes. It will seem very loose and wet.

4. Fold the dough from the sides to the center with a spatula until it forms a loose ball, usually 20 folds. You can also throw in a few stretch and folds here if you'd like.

Bulk Fermentation (10 to 18 hours)

Cover the dough and let it rest at 60–70°F (16–21°C) for 10 to 18 hours until the dough is at least doubled, but preferably tripled or more in volume, smells like bread, and jiggles nicely in its bowl when you shake it.

DAY 1 OR 2

Bake the Loaf (1 to 1½ hours)

1. Begin preheating the oven to 475°F (246°C).

2. Turn out the dough onto a generously floured sheet of parchment paper. With floured hands, fold two sides in to the center, then fold the other two sides in, forming a rustic boule. (The messy-looking folds should stay on top.)

3. Transfer the dough—parchment paper and all—to a room-temperature covered baker, cover, and let it rest until the oven preheats.

4. Bake 35 minutes covered, then uncover and bake until the interior is 207°F (97°C) and the crust browns to your liking, usually 35 to 50 minutes total time.

5. Cure in the cooling oven with the door ajar for 5 to 10 minutes. Cool 1 to 2 hours on a rack.

Part Four
Traditional Breads Using Your Sourdough Starter

CHAPTER 10

Mangia Bene, Ridi Spesso, Ama Molto: Recipes for Focaccia, Pizza, and Ciabatta

The breads in the following recipes would usually be made with commercial yeast. But when you've got a good starter that raises bread well and you've learned how to manage sourness, then there is absolutely no reason not to use your starter to make any yeasted bread you want. You may conclude, as I have, that these wild-yeast recipes are actually easier than most commercially yeasted recipes for the same breads. You might never buy yeast again!

Mangia bene, ridi spesso, ama molto means "eat well, laugh often, love much" in Italian. When you gather your tribe for a pizza party, there is destined to be good eating, frequent laughter, and a lot of love. Pizza making can sometimes feel intimidating for the baker: You have to figure out how to stretch crusts, deal with the toppings, and get it into a blisteringly hot oven and out again without burning your hands—it's an adventure. So I've started off the chapter with Focaccia (page 209), which is a much easier endeavor that can be topped with fun stuff too, is way simpler to prepare, and can be eaten warm (but will also be great later on). I'm not trying to dissuade you from the pizza party—I am committed to making it easy and awesome! After you've made focaccia, you'll be farther along the learning curve on stretching dough and baking topped bread.

When pizza day arrives, you'll be able to relax and enjoy the fun along with everyone else. I suggest starting with topping your pizza lightly and simply. This allows you to see what's going on with the crust while it's baking. You can even bake a crust with just tomato sauce and freeze it to make your own "frozen pizza" to top, bake, and enjoy later on. A common rookie mistake is to load up the pizza with too much sauce and toppings, which then prevents the dough from being able to bake properly. For rapid success, bake the first pizza with light sauce and toppings, then try adding more to the next pizzas and see how they turn out. It will be apparent where the "sweet spot" is when you bite into a slice.

This chapter includes two recipes for crust: the fermented Slow Wild Pizza Crust (page 212), and a faster version, Wild Pizza Crust Pronto (page 216), for pizza emergencies when you really cannot wait until tomorrow for pizza (we've all been there!). It also presents recipes for homemade toppings and a few winning combinations of toppings, just to get your creativity started.

When you're ready for more adventure, try the recipe for Ciabatta (page 225). This is a bread the Italians invented to entice Italian customers who were otherwise buying French baguettes to make sandwiches. Ciabatta is so delicious, the scheme worked—Italian panini are now enjoyed far and wide. The only challenge is the extremely wet dough (if you've tried the recipe for The Bomb (page 197), it will be familiar, only more wet). Once you get the hang of making this dough, you'll find it's a simple recipe for a luscious bread.

Buon appetito!

Focaccia

Focaccia is great for an appetizer, a snack, a road trip, or making interesting sandwiches. You can whip up this dough and then really express yourself with the toppings. Garlic and rosemary, radicchio and gorgonzola, olives and lemon zest, cherry tomatoes and thyme, sea salt and sesame, pine nuts and basil, figs and hazelnuts . . . these are just a few ideas to get your inspiration flowing. You can even go all out and put cheese on top like a pizza.

Spelt flour adds flavor depth and makes the dough easier to stretch, but you could substitute another whole wheat flour if you wish. This is a very easy bread to make.

MAKES 2 SMALL FOCACCIA OR 1 LARGE

TOTAL FLOUR: 450 GRAMS (1 POUND)

PRE-FERMENTED FLOUR: 17%

HYDRATION: 75%

WHOLE GRAIN: 20%

DOUGH: 797 GRAMS (1.8 POUNDS)

DAY 1

Build the Levain (10 minutes)

1. In a small bowl or pint jar, mix together:

30 grams starter that floats (2 tablespoons)

60 grams water (¼ cup)

60 grams whole spelt flour (½ cup)

2. Cover and let it rest at 60–70°F (16–21°C) for 8 to 12 hours until it is risen and bubbly and can float in water.

DAY 2

Mix the Dough (10 minutes)

1. Mix the flour in a large heavy bowl and set aside:

30 grams whole spelt flour (¼ cup)

345 grams bread flour (2⅞ cups)

2. Whisk together in the levain bowl or jar until uniform:

263 grams water (1⅛ cups) at 90°F (32°C)

150 grams levain (all)

3. Mix the wet ingredients into the flour, combining completely.

Autolyse (30 to 60 minutes)

1. Rest the dough for 30 to 60 minutes until it has relaxed in the bowl.

2. Add:

9 grams salt (1½ teaspoons)

continued

3. Sprinkle the salt over the wet dough and mix it in by poking and cutting it into the dough with a wet spatula, then folding the edges of the dough over the center with a spatula until the dough resists folding across itself, about 20 folds. Flip the ball of dough in the bowl and cover.

Bulk Fermentation (3 to 5 hours)

1. Rest the dough at 74–80°F (23–27°C), with two or three sets of stretch and folds 15 to 30 minutes apart until the dough holds its ball shape somewhat and makes a nice windowpane.
2. Let it rise until the dough is airy and puffy, double or triple or more in size, very jiggly, and smells like bread (3 to 5 hours total time after mixing in the levain).

Shape the Focaccia (5 to 15 minutes)

1. Line two 9-inch (23-centimeter) round or 8-inch (20-centimeter) square cake pans or 13-by-9-inch (33-by-23-centimeter) quarter baking sheet with parchment paper and coat with olive oil.
2. Turn out the dough onto a lightly misted counter and divide the dough if using two pans. Place the dough in the pans and stretch it gently to cover the entire pan. Aim for a consistent thickness so the dough bakes uniformly. It may need to rest 10 minutes to relax enough to stretch.

Proof the Focaccia (30 minutes to 18 hours)

Cover and retard the dough in the refrigerator for 4 to 18 hours. Or let it rest at 74–80°F (23–27°C) until risen and puffy (30 to 90 minutes).

DAY 2 OR 3

Top and Bake the Focaccia

1. Preheat the oven to 450°F (232°C) for 30 minutes. Place a rack in a high middle slot.
2. Dimple the top of the dough with oiled fingers, poking all the way to the bottom. Place toppings (such as herbs, tomatoes, olives, salt) onto the focaccia and poke the larger items into the dough. (Cheeses should be added 5 or 10 minutes before baking is finished.) Drizzle olive oil over the focaccia.
3. Bake on the rack for 25 to 30 minutes until the crust is deep golden brown. Check halfway through to see if the pan needs to be rotated or moved up or down in the oven for even baking.
4. Cool briefly and serve warm.

Slow Wild Pizza Crust

Time does the work while you dream up your toppings and start thinking of who to invite over for the pizza party. This recipe makes a thin crust pizza that is crisp on the bottom but flexible and tender to bite. The shaping directions ensure that the outer crust will have an open crumb.

The long, slow fermentation allows for maximum flavor development for a truly delicious pizza. The flours used in this recipe yield a tender crust, but any whole wheat can be used in place of durum, and all-purpose flour can be used in place of Tipo 00. The malted barley adds flavor and speeds rising, but it's not required.

MAKES FOUR 12- TO 14-INCH (30- TO 36-CENTIMETER) PIZZAS

FLOUR: 600 GRAMS (1.3 POUNDS)

PRE-FERMENTED FLOUR: 5%

HYDRATION: 70%

WHOLE GRAIN: 12%

DOUGH: 1032 GRAMS (2.3 POUNDS)

DAY 1

Mix the Dough (10 minutes)

1. Measure the flour into a large heavy bowl and set aside:

508 grams Tipo 00 flour (4¼ cups)

60 grams whole durum flour (½ cup)

12 grams malted barley flour (2 tablespoons)

12 grams salt (1¾ teaspoons)

2. Whisk the wet ingredients together in a small bowl until uniform:

400 grams water (scant 1¾ cups)

40 grams starter that floats (scant 3 tablespoons)

3. Mix the wet ingredients into the flour, combining completely. The dry flour needs to be fully incorporated. The dough will seem stiff and dry. If you're having trouble mixing, let it rest 10 minutes or so, then finish mixing.

Bulk Fermentation (8 to 18 hours)

1. Cover the dough and let it rest at 60–70°F (16–21°C) for 30 to 60 minutes. Then fold the dough with a wet spatula until it forms a loose ball and flip the ball.

2. Continue resting for 10 to 18 hours total until the dough is about doubled in volume.

continued

DAY 1 OR 2

Shape the Balls (10 minutes)

1. Turn the dough out onto a floured counter. It should seem puffy, sticky, and bubbly.

2. Divide the dough in four with a floured bench knife. Pre-shape each piece of dough: fold the dough into thirds, like a letter, fold in the other two sides, and flip the ball. Rotate it to tighten the skin. Let it rest a couple of minutes.

3. Shape into balls again with a nice taut skin and seal the bottoms well.

4. If you don't want to bake the pizzas right away, prepare four bowls that hold about 2 cups, such as soup bowls, with ½ teaspoon olive oil in each one. Place the balls in the olive-oiled bowls seam side down, rolling in the oil to coat them well. Cover the bowls with plates and refrigerate up to 24 hours. Continue with the next step when ready.

5. Rest for 10 to 60 minutes before making pies so the gluten can relax.

DAY 2 OR 3

Stretch into Crusts (20 minutes)

1. Preheat the oven to 550°F (288°C) with a baking stone or an inverted baking sheet on a low rack for 1 hour. While the oven is heating you can assemble the toppings and stretch the dough into crusts.

For each pizza:

2. Turn out the ball of dough seam side up onto a generously floured counter. The top of the dough ball will become the bottom of the pizza and the sticky seam side will get the toppings.

3. Being careful not to deflate it around the edges to preserve a puffy edge, or cornicione, stretch the dough, seam side up, into a 12- to 14-inch (30- to 36-centimeter) round and place it on parchment paper.

Here's my method, although there are many good ways to do this. With floured hands I hold the dough up like a steering wheel (sticky side toward me) and turn the wheel around and around until it's stretched to about 8 inches (20 centimeters) in diameter. Then I drape it on the backs of my floured hands (sticky side up) and move my hands around and around under the edges of the dough, stretching it by gravity but being careful not to make any holes. Then I lay it, sticky side up, on parchment paper. It always shrinks a little after I lay it down, so I stretch it as big as I can manage.

If you're having a lot of trouble, flour and a rolling pin are okay to use, too (pizza should be fun; however, large bubbles will be rolled out of the dough, resulting in a denser crust. If the dough is fighting you and springing back into a small disk, it needs to rest. Turn your attention to a different pizza and come back to it in 10 minutes.

4. The stretched crusts can be allowed to proof for 20 to 30 minutes prior to topping if desired or topped and baked immediately.

Top and Bake (1 hour)

Top the pizza at the very last minute, then bake it immediately to avoid a soggy crust.

For each pizza:

1. Top with sauce and a single layer of toppings, then cheese.

2. Immediately transfer the parchment paper and pizza to the stone or sheet.

3. Bake for 6 to 12 minutes until the edges of the crusts are browned to your liking and toppings are cooked. You may need to rotate the pizza for even cooking. If the top cooks before the crust, lower the stone, possibly onto the floor of the oven.

Wild Pizza Crust Pronto

If the mood for pizza strikes unexpectedly, this recipe will get you to pizza dough that's ready to shape (or store in the refrigerator) in about 4 hours with very little hands-on effort. You can save your energy for the pizza party. This dough will make fantastic pizza with a thin crust that features a crispy bottom, a chewy center, and big open bubbles in the edges. The smell of this baking—oh my!

We add in some stretch and folds, and use the ideal temperature for yeast to speed along the dough development. This recipe works best with a recently fed starter. The amount called for uses all your starter if you keep it at 60 grams, but don't worry. Your starter will easily regrow from the scraps on the walls of the starter jar. Spelt gives the dough extra stretch for easy shaping, but any whole wheat can be substituted. Also, all-purpose flour can be used in place of Tipo 00. The malted barley adds flavor and speeds rising, but it's not required.

MAKES FOUR 12- TO 14-INCH (30- TO 36-CENTIMETER) PIZZAS

FLOUR: 600 GRAMS (1.3 POUNDS)

PRE-FERMENTED FLOUR: 5%

HYDRATION: 65%

WHOLE GRAIN: 12%

DOUGH: 1002 GRAMS (2.2 POUNDS)

DAY 1

Mix the Dough (10 minutes)

1. Mix the flours together in a large heavy bowl and set aside:

498 grams Tipo 00 flour (4⅛ cups)

60 grams whole spelt flour (½ cup)

12 grams malted barley flour (2 tablespoons)

2. Whisk the wet ingredients together in a small bowl until uniform:

360 grams water (scant 1⅛ cups) at 90°F (32°C)

60 grams starter that floats (¼ cup)

3. Mix the wet ingredients into the flour, combining completely. The dry flour needs to be fully incorporated. The dough will seem stiff and dry. If you're having trouble mixing, let it rest 10 minutes or so, then finish mixing.

Autolyse (30 to 60 minutes)

1. Rest the dough for 30 to 60 minutes until it has relaxed in the bowl.

2. Add:

12 grams salt (1¾ teaspoons)

3. Sprinkle the salt over the dough and mix it in by poking and cutting it into the dough with a wet spatula, then fold the edges of the dough over the center with a spatula or wet hands until the dough resists folding across itself, about 20 folds. Flip the ball of dough in the bowl and cover.

Bulk Fermentation (3 to 5 hours)

1. Rest the dough at 74–80°F (23–27°C), with two to four sets of stretch and folds 15 to 30 minutes apart until the dough holds its ball shape somewhat and makes a nice windowpane.

2. Let it rise until the dough is bubbly and puffy, double in size, not quite jiggly, and smells like dough (3 to 5 hours total time after mixing in the starter).

Shape the Balls (10 minutes)

1. Turn the dough out onto a floured counter. It should seem puffy, sticky, and bubbly.

2. Divide the dough in four with a floured bench knife. Pre-shape each piece of dough: fold the dough into thirds, like a letter, fold in the other two sides, and flip the ball. Rotate it to tighten the skin. Let it rest a couple of minutes.

3. Shape into balls again with a nice taut skin and seal the bottoms well.

continued

4. If you don't want to bake the pizzas right away, prepare four bowls that hold about 2 cups, such as soup bowls, with ½ teaspoon olive oil in each one. Place the balls in the olive-oiled bowls seam side down, rolling in the oil to coat them well. Cover the bowls with plates and refrigerate up to 24 hours. Continue with the next step when ready.

5. Rest for 10 to 60 minutes before making pies so the gluten can relax.

DAY 1 OR 2

Stretch into Crusts

1. Preheat the oven to 550°F (288°C) with a baking stone or an inverted baking sheet on a low rack for 1 hour. While the oven is heating you can assemble the toppings and stretch the dough into crusts.

For each pizza:

2. Turn out the ball of dough seam side up onto a generously floured counter. The top of the dough ball will become the bottom of the pizza and the sticky seam side will get the toppings.

3. Being careful not to deflate it around the edges to preserve a puffy edge, or cornicione, stretch the dough, seam side up, into a 12- to 14-inch (30- to 36-centimeter) round and place it on parchment paper.

Here's my method, although there are many good ways to do this. With floured hands I hold the dough up like a steering wheel (sticky side toward me) and turn the wheel around and around until it's stretched to about 8 inches (20 centimeters) in diameter. Then I drape it on the backs of my floured hands (sticky side up) and move my hands around and around under the edges of the dough, stretching it by gravity but being careful not to make any holes. Then I lay it, sticky side up, on parchment paper. It always shrinks a little after I lay it down, so I stretch it as big as I can manage.

If you're having a lot of trouble, flour and a rolling pin are okay to use, too (pizza should be fun; however, large bubbles will be rolled out of the dough, resulting in a denser crust. If the dough is fighting you and springing back into a small disk, it needs to rest. Turn your attention to a different pizza and come back to it in 10 minutes.

4. The stretched crusts can be allowed to proof for 20 to 30 minutes prior to topping if desired or topped and baked immediately.

Top and Bake (1 hour)

Top the pizza at the very last minute, then bake it immediately to avoid a soggy crust.

For each pizza:

1. Top with sauce and a single layer of toppings, then cheese.

2. Immediately transfer the parchment paper and pizza to the stone or sheet.

3. Bake for 6 to 12 minutes until the edges of the crusts are browned to your liking and toppings are cooked. You may need to rotate the pizza for even cooking. If the top cooks before the crust, lower the stone, possibly onto the floor of the oven.

The Newman Special Pizza

This is one pizza everyone in my family loves. We love it so much we have dubbed it The Newman Special. There is something about the combination of flavors that is just so satisfyingly delicious. The Italian sausage can be browned in the oven while it's preheating.

MAKES ONE PIZZA

1 stretched 12- to 14-inch Slow Wild Pizza Crust (page 212, or Wild Pizza Crust Pronto, page 216; 30 to 36 centimeters), ready to be topped

⅓ cup Summer Roma Tomato Sauce (recipe follows; 75 grams)

⅓ cup Italian sausage, uncased, browned, and drained (50 grams)

⅓ cup artichoke hearts, drained of water and cut to bite-sized pieces (87 grams)

1 cup fresh baby spinach leaves (30 grams)

2 ounces grated mixed Italian cheeses (such as mozzarella, Asiago, provolone, Parmesan; 57 grams)

Assemble the Pizza and Bake (1 hour)

1. Preheat the oven to 550°F (288°C) with a baking stone or an inverted baking sheet on a low rack for 1 hour.

2. Spread the tomato sauce around the pizza, leaving a gap near the edge. I like to use a gravy spoon for this.

3. Sprinkle the Italian sausage, then the artichoke hearts, followed by the spinach. Top with the cheese.

4. Immediately transfer the parchment paper and pizza to the stone or sheet.

5. Bake for 6 to 12 minutes until the edges of the crusts are browned to your liking and toppings are cooked. You may need to rotate the pizza for even cooking. If the top cooks before the crust, lower the stone, possibly onto the floor of the oven.

continued

Summer Roma Tomato Sauce

Roma tomatoes, those oblong gems also known as paste tomatoes or Italian plum tomatoes, make excellent sauce because they have very little water compared to tomatoes bred for fresh eating. For making delicious sauce that's worth the trouble, the Romas must be in season, a deep, robust red color, and just beginning to soften. If you can get Romas or San Marzanos from your garden or farmers' market, all the better. In this recipe, we boil off much of the water, leaving behind a concentrated tomato flavor that sings on a pizza or pasta. It takes about an hour of light attention, so pour yourself a nice glass of wine to sip while you stir.

This recipe scales up nicely if you manage to secure a bushelful (I usually make a double recipe in an 8-quart kettle). Extra sauce can be canned or frozen for use over the winter. Rather than laboriously seeding and skinning my tomatoes, I keep it simple and puree them whole. If you like, you can run the finished sauce through a food mill to remove the seeds. For faster sauce, use a wider pan.

MAKES 2 QUARTS (1.9 LITERS)

6 pounds fresh, red, ripe Roma tomatoes (2.7 kilograms)

4 teaspoons salt (24 grams)

Dried herbs such as oregano, thyme, and marjoram to taste (optional)

Make the Sauce

1. Remove the stem end of the tomatoes and quarter them.
2. Working in batches, process them in your food processor for about 15 seconds, or until they liquify, then transfer them to a roomy and heavy-bottomed kettle.
3. Place the kettle over high heat and do not cover. We are going to boil off a lot of water from the tomatoes while watching them to prevent boiling over (likely near the beginning) or sticking on the bottom (likely near the end). Stay nearby and stir the tomatoes periodically until the sauce thickens up. This usually takes 30 to 50 minutes.
4. Grind the tomatoes again for about 30 seconds (an immersion blender is handy for this step) and add the salt and the herbs, if using.
5. Continue cooking over high heat until it looks smooth and thick and tastes delicious, 45 minutes to an hour of total cooking time.
6. Pack into pint or quart containers and refrigerate or freeze if not using immediately. This sauce can also be canned using the U.S.D.A. guidelines, which can be found on the internet at https://nchfp.uga.edu/publications/publications_usda.html.

Roasted Vegetable Pizza

Elevate the vegetarian pizza to the next level. Roasting concentrates, deepens, and sweetens the flavor of vegetables and reduces the amount of water they release on your pizza. Top choices include mushrooms, red bell peppers, onions, zucchini, fennel bulb, radicchio, asparagus (slice diagonally), fresh corn kernels, elephant garlic, and Romanesco. I recommend choosing one or two different vegetables per pizza. I usually roast the vegetables in the oven while it's preheating for the pizza.

MAKES ONE PIZZA

1 cup raw vegetables sliced ⅛ inch thick (3 millimeters)

salt to taste

⅓ cup Summer Roma Tomato Sauce (page 220; 75 grams)

1 stretched 12- to 14-inch Slow Wild Pizza Crust (page 212, or Wild Pizza Crust Pronto, page 216; 30 to 36 centimeters), ready to be topped

2 ounces grated mixed Italian cheeses (such as mozzarella, Asiago, provolone, Parmesan; 57 grams)

Roast the Vegetables

1. Preheat the oven to 475°F (246°C) with a rack in the center and a baking stone or an inverted baking sheet on a low rack.

2. Lightly oil a baking sheet and arrange the vegetables in a single layer. Sprinkle with salt.

3. Roast in the oven 5 to 10 minutes until they are a little golden on the tops or the bottoms. Check them frequently so they don't burn. Remove them from the pan promptly to prevent over-browning.

Assemble the Pizza and Bake (1 hour)

1. Remove the center rack from the oven and raise the temperature to 550°F (288°C). Allow the oven to preheat for another 15 to 30 minutes.

2. Spread the tomato sauce around the pizza, leaving a gap near the edge. I like to use a gravy spoon for this.

3. Sprinkle the roasted vegetables around the pizza and top with the cheese.

4. Immediately transfer the parchment paper and pizza to the stone or sheet.

5. Bake for 6 to 12 minutes until the edges of the crusts are browned to your liking and toppings are cooked. You may need to rotate the pizza for even cooking. If the top cooks before the crust, lower the stone, possibly onto the floor of the oven.

Margherita Pizza

This pizza is a classic for good reason. The delicate sweetness of the fresh mozzarella pairs with the heavenly aroma of basil in a delightful combination. If you prefer, the basil can also be placed on the pizza after it's baked.

MAKES ONE PIZZA

1 stretched 12- to 14-inch Slow Wild Pizza Crust (page 212, or Wild Pizza Crust Pronto, page 216; 30 to 36 centimeters), ready to be topped

⅓ cup Summer Roma Tomato Sauce (page 220; 75 grams)

6 to 12 slices fresh tomato, ⅛ inch (3 millimeters) thick

6 to 12 fresh basil leaves

4 ounces fresh mozzarella, drained and sliced ¼ inch (6 millimeters) thick or torn into shreds (113 grams)

Assemble the Pizza and Bake (1 hour)

1. Preheat the oven to 550°F (288°C) with a baking stone or an inverted baking sheet on a low rack for 1 hour.
2. Spread the tomato sauce around the pizza, leaving a gap near the edge. I like to use a gravy spoon for this.
3. Arrange the tomato slices around the pizza.
4. Press the basil leaves into the sauce and top with the cheese.
5. Immediately transfer the parchment paper and pizza to the stone or sheet.
6. Bake for 6 to 12 minutes until the edges of the crusts are browned to your liking and toppings are cooked. You may need to rotate the pizza for even cooking. If the top cooks before the crust, lower the stone, possibly onto the floor of the oven.

Pesto Pizza

Green pizza, yes! Pesto shines on this simply luscious pizza.

MAKES ONE PIZZA

1 stretched 12- to 14-inch Slow Wild Pizza Crust (page 212, or Wild Pizza Crust Pronto, page 216; 30 to 36 centimeters), ready to be topped

⅓ cup Pesto alla Genovese (recipe follows; 84 grams)

12 to 18 cherry tomatoes, halved

2 ounces grated mixed Italian cheeses (such as mozzarella, Asiago, provolone, Parmesan; 57 grams)

1 teaspoon pine nuts (3 grams)

Assemble the Pizza and Bake (1 hour)

1. Preheat the oven to 550°F (288°C) with a baking stone or an inverted baking sheet on a low rack for 1 hour.

2. Spread the pesto around the pizza, leaving a gap near the edge. I like to use an offset spatula for this.

3. Arrange the cherry tomatoes around the pizza.

4. Top with the cheese and sprinkle with the pine nuts.

5. Immediately transfer the parchment paper and pizza to the stone or sheet.

6. Bake for 6 to 12 minutes until the edges of the crusts are browned to your liking and toppings are cooked. You may need to rotate the pizza for even cooking. If the top cooks before the crust, lower the stone, possibly onto the floor of the oven.

continued

Pesto alla Genovese

Perfect for topping a pizza, this pesto can also be used to dress pasta or vegetables, dropped into a bowl of vegetable soup, or spread in panini. When fresh basil is in season, you can make a big batch, then freeze the extra in small containers to use over the winter.

Traditionally ground in a mortar and pestle, pesto is easy to whip up in a food processor. If omitting the cheese, which I prefer to add just before serving pesto, add ½ teaspoon (3 grams) salt. To prevent the pesto from turning brown when exposed to air, a couple of tablespoons of lemon juice can be added to the pesto during grinding, and then a layer of olive oil can be poured over the finished pesto to protect it from oxygen.

MAKES ABOUT 2 CUPS (473 MILLILITERS) PESTO

1 to 3 garlic cloves

1 cup pine nuts or walnuts (150 grams)

3 cups fresh basil leaves, de-stemmed and packed (about 120 grams)

¾ cup extra virgin olive oil (100 grams), chilled

¼ cup fresh grated Parmigiano-Reggiano cheese (25 grams)

¼ cup fresh grated Pecorino Romano cheese (25 grams)

Make the Pesto

1. Grind the garlic and nuts briefly in a food processor.
2. Add the basil and grind in pulses so as not to overheat, pausing to stir if needed, until uniformly shredded.
3. Drizzle the olive oil in slowly while the machine is running and process until smooth, watching that it doesn't become warm.
4. Add the cheese or salt and process just until mixed.
5. Pack into small jars and refrigerate or freeze if not using immediately.

Ciabatta

Crusty with a soft interior and honeycombed with giant holes, ciabatta is a top-notch bread for sopping up olive oil or sauces, dunking in soup, or making crostini for bruschetta. When sliced horizontally, it makes fantastic sandwiches or panini. Very high hydration, extra stretch and folds, minimal handling during shaping, and long, cold fermentation combine to create this bread, which seems almost universally adored.

This is not a difficult recipe, but I do not recommend it for beginners. The dough is so wet, loose, and sticky it will seem insane. But as you go through the folds, the gluten will build, reducing stickiness and developing strength for an open crumb. A generous use of flour while shaping and minimal touching help you get this dough into the oven rather than all over yourself. Plan to bake on a baking sheet or stone with a steam pan.

MAKES 2 CIABATTAS

TOTAL FLOUR: 450 GRAMS (1 POUND)
PRE-FERMENTED FLOUR: 3%
HYDRATION: 85%
WHOLE GRAIN: 0%
DOUGH: 842 GRAMS (1.9 POUNDS)

DAY 1

Mix the Dough (10 minutes)

1. Measure the flour into a large heavy bowl and set aside:

435 grams all-purpose flour (3⅝ cups)

9 grams salt (1½ teaspoons)

2. Whisk the wet ingredients together in a small bowl until uniform:

368 grams water (1½ cups plus 1 tablespoon)

30 grams starter that floats (2 tablespoons)

3. Mix the wet ingredients into the flour, combining completely. Stir vigorously until the dough is uniform. It will seem very shaggy and wet like a batter and will not form a ball but do not add more flour.

Bulk Fermentation (14 to 32 hours)

1. Cover the dough and let it rest at 60–70°F (16–21°C) for 30 to 60 minutes. Then fold the dough with a wet spatula or wet hands until it forms a loose, very flattened ball and flip the ball.

2. Repeat the rest and stretch-and-fold process at least three times more, with 15 to 30 minutes of rest in between to allow the gluten to relax. For this very wet dough, I prefer coil folding. By the final set of folds, you should be able to slowly lift the dough out of the bowl without it sticking and the dough should feel strong and bouncy.

continued

3. Continue resting until the dough has doubled or tripled in size and is jiggly and smells ready, with many large bubbles visible on the surface, usually 6 to 8 hours since mixing.

4. Move the dough to the refrigerator for 12 to 24 hours.

DAY 2

Shape the Loaf (10 to 20 minutes)

1. Bring the dough out of the refrigerator and prepare a parchment paper–lined peel or baking sheet for resting the ciabattas.

2. Turn the dough out in one piece onto a generously floured counter. It should spread out all wet and bubbly. Handle minimally to achieve an open crumb and use plenty of flour on the outside of the ciabattas but dust off any that will end up in the interior while you are shaping.

3. Gently tuck your floured hands or bench knife under the dough to coax it into a rough rectangle, stretching it without deflating it until it is about 10 by 14 inches (25 by 36 centimeters) and an even thickness throughout. If the dough resists, let it rest and warm up on the counter for 10 minutes.

4. Fold the dough in half so that it is about 5 by 14 inches (13 by 36 centimeters). Sprinkle with flour.

5. Dimple it with your fingers for the iconic flattened ciabatta ("slipper") shape. Cut it into two 5-by-7-inch (13-by-18-centimeter) pieces with a floured bench knife.

6. Carefully lift each piece from underneath and stretch it to 10 inches (25 centimeters) long as you invert it onto the parchment paper, leaving space between the ciabattas. Gently nudge them to correct their shape if necessary. They will seem a little like bags of water.

7. Sprinkle with more flour so they are fully coated. Cover with parchment paper, an overturned roasting pan, or a puffed-up bag to prevent drying.

Proof the Ciabattas (30 to 60 minutes)

Rest the loaves at room temperature for 30 to 60 minutes just until the ciabattas seem risen and jiggly.

Bake the Ciabattas (1 to 1½ hours)

1. Preheat the oven to 475°F (246°C) 30 to 60 minutes before baking, and prepare your baking and steam setup.

2. Transfer the ciabattas and their parchment paper to the baking stone or place the baking sheet in the oven.

3. Add 1½ cups (355 milliliters) of water to the steam pan.

4. Bake 15 to 20 minutes with steam, until they seem finished rising, then bake until the interior is 207°F (97°C) and the crust browns to your liking, usually 20 to 30 minutes total time for a golden to medium brown. Rearrange the ciabattas as needed for even browning of the crusts during the final 10 minutes.

5. Cure in the cooling oven with the door ajar 5 to 10 minutes. Cool 1 to 2 hours on a rack.

CHAPTER 11

Jewish Breads: Bagels, Challah, and Deli Rye

Perhaps it's not surprising that a people for whom bread fell from heaven came up with celestially delicious breads. This chapter presents recipes for my personal favorites. These recipes differ markedly from those in the previous chapters. Here, the goal is not to create a lofty crumb with open holes. These breads are meant to be substantial, with a chewy, velvety crumb and a supple but sturdy crust. You'll find that the hydration is much lower, the dough is stiffer and handles differently, and baking temperatures are cooler. If you don't have bread flour, you can substitute all-purpose flour. Your breads will be just a bit softer and less chewy. Or, refer to Chapter 3 for how to strengthen all-purpose flour with vital wheat gluten to create a substitute for bread flour.

Maybe you don't think of sourdough when you think of challah or bagels, but you can make these with your sourdough starter and they will taste amazing—not necessarily sour, just really good. In these recipes we manage the fermentation to keep the dough "sweet." Imagine the luxurious joy of spreading cream cheese on a still-warm bagel on a weekend morning. And shaping bagels can be a fun way to enjoy cooking together, since anyone can enjoy rolling dough into ropes! You'll be presented with options to make dense chewy bagels or light fluffy bagels, so you can perfect them as you like. Break into a warm, fresh challah to pass around the table that has full-bodied flavor, even when eaten plain. Pile pastrami on a deli rye that has a touch of sourness, as it should. Did you know Jewish rye breads made with a commercial yeast often have acid additives to mimic the flavors that the bacteria in your starter naturally impart? For this reason, and because it may be hard to find satisfying versions of these breads where you live, it is a worthy endeavor to make them at home with your very own wild yeast.

Slow Bagels

Once you have eaten your own fresh-baked sourdough bagels, you may never look back. They are so easy and fun to make. You mix the dough the night before and retard it in bulk fermentation. The dough does not really need to rise much for bagels, but the flavor really develops during a slow fermentation. After shaping, the bagels are kettled to create the distinctive bagel crust and chewiness.

The dough is a lower hydration, and during mixing it will feel somewhat stiffer than most recipes in this cookbook. Allowing the dough to rest, then finishing up the mixing process with some folds, makes things much easier.

MAKES 8 BAGELS ABOUT 3½ INCHES (9 CENTIMETERS) IN DIAMETER

TOTAL FLOUR: 450 GRAMS (1 POUND)

PRE-FERMENTED FLOUR: 3%

HYDRATION: 60%

WHOLE GRAIN: 10%

DOUGH: 729 GRAMS (1.6 POUND)

DAY 1

Mix the Dough (10 minutes)

1. Mix in a large heavy bowl or in the bowl of your stand mixer:

390 grams bread flour (3¼ cups)

36 grams whole wheat flour (¼ cup plus 1 tablespoon)

9 grams diastatic barley malt powder (1 tablespoon)

9 grams salt (1½ teaspoons)

2. Whisk the wet ingredients together in a small bowl until uniform:

255 grams water (1 cup plus 1 tablespoon)

30 grams starter that floats (2 tablespoons)

3. Mix the wet ingredients into the flour and combine completely with a silicone spatula or mixer until all the flour is wet and it forms a ball. The dough will feel very stiff and dry. This usually takes 2 to 3 minutes. Toward the end, you can use a floured hand to finish mixing, if needed. Do not add additional flour or water. If it is too hard to incorporate all the flour, cover and let it rest for 15 minutes or longer, then finish mixing.

continued

Bulk Fermentation (8 to 24 hours)

1. Cover the bowl with a lid or plate and let the dough rest for 1 to 2 hours at room temperature, until it is relaxed and softened.

2. Fold the dough in the bowl until it is smooth and forms a stiff ball, 10 to 15 folds. Flip the ball and cover.

3. Rest the dough for 8 to 24 hours total, starting at 60–70°F (16–21°C) for 4 to 12 hours, then refrigerate it until you are ready to use it.

DAY 1 OR 2

Shape the Bagels (20 minutes)

1. Turn the dough out onto a clean counter and divide it into eight pieces with a bench knife. Each piece should weigh around 90 grams (3¼ ounces).

2. Pre-shape each piece: fold the dough into thirds, like a letter, fold in the other two sides, and flip the ball. Rotate it to tighten the skin. Let it rest 10 to 15 minutes. For bagels with a close, chewy crumb, firmly press the gas out of the dough while pre-shaping. If you like fluffy bagels, preserve the loft.

3. Shaping option 1: Shape each piece into a taut boule and tighten the skin, resting to seal the bottom while you shape the others. Poke a finger into the center of each ball until it comes through the other side, place another finger from your other hand in the hole, and rotate them until the hole is 2 inches (5 centimeters) in diameter. The hole should seem overly large because it will become much smaller in the finished bagel.

4. Shaping option 2: Roll each ball of dough into a rope, draping it around your hand and carefully pinching the overlapping ends together to form a ring. The hole will seem large, but it will become much smaller in the finished bagel.

Proof the Bagels (10 to 30 minutes)

Rest the bagels at room temperature on a floured or oiled surface until they are just a bit risen. I like to use oiled parchment paper on a baking sheet so I can transport the bagels to the kettle easily.

Kettle and Bake the Bagels (1 hour)

1. Preheat the oven to 375°F (191°C).

2. In a saucepan, heat to a boil:

> **2 liters (2 quarts) water (should be 4 inches [10 centimeters] deep in the saucepan)**
>
> **40 grams barley malt syrup, or substitute honey or brown sugar (2 tablespoons)**
>
> **15 grams baking soda (1 tablespoon)**

3. Line an 18-by-13-inch (46-by-33-centimeter) baking sheet with a piece of parchment paper, oiled or sprinkled with cornmeal.

4. Lower the bagels into the boiling water, a few at a time and boil for 30 seconds per side, or up to 2 minutes per side for denser, chewy bagels. Drain the bagels well (so they don't stick) and place them, top side up, on the oiled parchment paper. Continue with the next batch until all the bagels are boiled.

5. Top the wet bagels by sprinkling with (or dipping them into a bowl of) sesame or poppy seeds, dried onions or garlic, or a pinch of coarse salt (optional).

6. Transfer the baking sheet to a middle rack in the oven and bake 30 to 45 minutes until the interior is 207°F (97°C) and golden to medium brown, checking for even baking halfway through and moving or rotating the baking sheet as needed. Add a second baking sheet below the bagels if the bottoms are browning too fast. Tent with foil or lower the rack if the tops are browning too fast, paying special attention not to burn toppings if used.

7. Cool for at least 30 minutes on a wire rack. Warm bagels are delicious, but hot bagels will still be doughy inside.

Whole Wheat Bagels

With their extra flavor, texture, and chewiness, these whole wheat bagels are great topped with some smoked fish, cream cheese, tomatoes, and cucumbers.

The optional spatula folds help develop the gluten for easier shaping and taller bagels.

MAKES 8 MEDIUM-SIZED BAGELS

TOTAL FLOUR: 450 GRAMS (1 POUND)

PRE-FERMENTED FLOUR: 3%

HYDRATION: 67%

WHOLE GRAIN: 50%

DOUGH: 759 GRAMS (1.7 POUNDS)

DAY 1

Mix the Dough (10 minutes)

1. Mix in a large heavy bowl:

> **210 grams bread flour (1¾ cups)**
>
> **217 grams whole wheat flour (1¾ cups plus 1 tablespoon)**
>
> **8 grams diastatic barley malt powder (1 tablespoon)**
>
> **9 grams salt (1½ teaspoons)**

2. Whisk the wet ingredients together in a small bowl until uniform :

> **285 grams water (1¼ cups)**
>
> **30 grams starter that floats (2 tablespoons)**

3. Mix the wet ingredients into the flour and combine completely with a silicone spatula until all the flour is wet and it forms a ball. The dough will feel very stiff and dry. This usually takes 2 to 3 minutes. Toward the end, you can use a floured hand to finish mixing, if needed. Do not add additional flour or water. If it is too hard to incorporate all the flour, cover and let it rest for 15 minutes or longer, then finish mixing.

Bulk Fermentation (8 to 24 hours)

1. Cover the bowl with a lid or plate and let the dough rest for 1 to 2 hours at room temperature, until it is relaxed and softened.

2. Fold the dough in the bowl until it is smooth and forms a stiff ball, 10 to 15 folds. Flip the ball and cover.

3. Rest the dough for 8 to 24 hours total, starting at 60–70°F (16–21°C) for 4 to 12 hours, then refrigerate it until you are ready to use it.

DAY 1 OR 2

Shape the Bagels (20 minutes)

1. Turn the dough out onto a clean counter and divide it into eight pieces with a bench knife. Each piece should weigh around 90 grams (3¼ ounces).

2. Pre-shape each piece: fold the dough into thirds, like a letter, fold in the other two sides, and flip the ball. Rotate it to tighten the skin. Let it rest 10 to 15 minutes. For bagels with a close, chewy crumb, firmly press the gas out of the dough while pre-shaping. If you like fluffy bagels, preserve the loft.

3. Shaping option 1: Shape each piece into a taut boule and tighten the skin, resting to seal the bottom while you shape the others. Poke a finger into the center of each ball until it comes through the other side, place another finger from your other hand in the hole, and rotate them until the hole is 2 inches (5 centimeter) in diameter. The hole should seem overly large because it will become much smaller in the finished bagel.

4. Shaping option 2: Roll each ball of dough into a rope, draping it around your hand and carefully pinching the overlapping ends together to form a ring. The hole will seem large, but it will become much smaller in the finished bagel.

Proof the Bagels (10 to 30 minutes)

Rest the bagels at room temperature on a floured or oiled surface until they are just a bit risen. I like to use oiled parchment paper on a baking sheet so I can transport the bagels to the kettle easily.

Kettle and Bake the Bagels (1 hour)

1. Preheat the oven to 375°F (191°C).

2. In a saucepan, heat to a boil:

    2 liters (2 quarts) water (should be 4 inches [10 centimeters] deep in the saucepan)

    40 grams barley malt syrup, or substitute honey or brown sugar (2 tablespoons)

    15 grams baking soda (1 tablespoon)

3. Line an 18-by-13-inch (46-by-33-centimeter) baking sheet with parchment paper, oiled or sprinkled with cornmeal.

4. Lower the bagels into the boiling water, a few at a time and boil for 30 seconds per side, or up to 2 minutes per side for denser, chewy bagels. Drain the bagels well (so they don't stick) and place them, top side up, on the oiled parchment paper. Continue with the next batch until all the bagels are boiled.

5. Top the wet bagels by sprinkling with (or dipping them into a bowl of) sesame or poppy seeds, dried onions or garlic, or a pinch of coarse salt (optional).

6. Transfer the baking sheet to a middle rack in the oven and bake 30 to 45 minutes until the interior is 207°F (97°C) and golden to medium brown, checking for even baking halfway through and moving or rotating the baking sheet as needed. Add a second baking sheet below the bagels if the bottoms are browning too fast. Tent with foil or lower the rack if the tops are browning too fast, paying special attention not to burn toppings if used.

7. Cool for at least 30 minutes on a wire rack. Warm bagels are delicious, but hot bagels will still be doughy inside.

Wild Challah

What is more beautiful than a braided bread? This egg-and-honey loaf is festive and delicious with a meal, and it makes amazing French toast or bread pudding if there's any left over. I adore challah and I especially like it when the crumb is chewy and stretchy and long strands of bread can be pulled from the loaf like string cheese. Follow the shaping instructions carefully if you want that effect in your challah.

This recipe uses a fresh levain and a fast bulk fermentation at the yeast's ideal temperature to tone down acids. The overnight retard adds all the complex flavors of a long fermentation but imperceptible sourness. If your starter runs sour, refresh it two or three times over the day or two prior to making the challah, and keep the starter below 78°F (26°C). Egg dough is an incredibly sticky beast until the gluten is fully developed. The added eggs and oil also inhibit gluten formation, so this dough needs kneading to develop the gluten. For these reasons, I prefer to mix the dough in a stand mixer and to knead it until it is satiny and easy to handle. If you want a big challah, double the recipe and make a four-strand braid.

MAKES ONE LOAF

TOTAL FLOUR: 450 GRAMS (1 POUND)

PRE-FERMENTED FLOUR: 17%

HYDRATION: 57%

WHOLE GRAIN: 0%

DOUGH: 797 GRAMS (1.8 POUNDS)

DAY 1

Build the Levain (10 minutes)

1. In a small bowl or pint jar, mix together:

> **30 grams starter that floats (2 tablespoons)**
>
> **60 grams water (¼ cup)**
>
> **60 grams all-purpose flour (½ cup)**

2. Cover and let it rest at 60–70°F (16–21°C) for 8 to 12 hours until it is risen and bubbly and can float in water.

DAY 2

Mix the Dough (15 minutes)

1. Mix the flours in a large heavy bowl or in the bowl of your stand mixer and set aside:

367 grams bread flour (3 cups)

9 grams diastatic barley malt powder (1 tablespoon)

2. Whisk together in a medium bowl until uniform:

3 large eggs, whisked (150 grams, ¾ cup)

32 grams water (2 tablespoons)

45 grams honey (2 tablespoons plus 1 teaspoon)

45 grams oil, such as sunflower (3 tablespoons plus 1 teaspoon)

150 grams ready levain (all, ⅔ cup)

3. Add the egg mixture to the flour and combine completely with a silicone spatula or a mixer until all the flour is wet and it forms a ball. This usually takes 2 to 3 minutes. The dough will feel very sticky and seem like a batter. Toward the end, you may prefer to use a damp hand to finish mixing if needed. Do not add additional flour or water.

Autolyse (30 minutes)

1. Cover the bowl with a lid or plate and let the dough rest for 30 minutes, until it is relaxed and softened.

2. Add:

9 grams salt (1½ teaspoons)

3. Sprinkle the salt evenly on the surface of the dough and knead, by hand or in a mixer on a low speed, until the dough becomes a bit firmer than an earlobe and satiny on the surface, about 10 minutes. If you measured by cups, adjust the flour or water if necessary and continue kneading to obtain the correct consistency: the dough should be firm but still a little soft and sticky, not stiff.

4. Form the dough into a ball and place in an oiled bowl. Cover with a lid or plate.

Bulk Fermentation (3 to 5 hours)

1. Rest the dough at 74–80°F (23–27°C). After 60 minutes, when the dough has relaxed, perform a set of stretch and folds, then flip the ball. Your dough should feel strong and elastic.

2. Continue resting the dough until it is about double its original volume and smells more like bread than flour and a finger poke fills back in slowly (3 to 5 hours total time after mixing in the levain).

Shape the Braid (20 minutes)

1. Turn the dough out onto a clean counter and stretch it into a small rectangle. Divide the dough into three long pieces with a bench knife. Each piece should weigh about 280 grams (10 ounces).

continued

2. Pre-shape each piece into a log by lifting it by the two ends, stretching then folding it in half, holding it up by its ends again to stretch it and fold it in half again. Then let the dough rest, covered, 10 to 15 minutes until it relaxes.

3. Shape each log into a long strand by stretching (not rolling) the dough, holding it up by one end and gently pulling down all along its length. If it becomes difficult to stretch, rest the dough for 10 minutes before continuing, until each strand is about 14 inches (36 centimeters) long.

4. Press the top ends of the strands together until they adhere and braid them to the bottom. For the best definition of the braid in the finished loaf, braid loosely. Press the ends together to adhere on the bottom end. Tuck the ends under the loaf or, for a round challah, bring the ends together and press to adhere them.

5. Place the braid on a parchment paper–lined baking sheet and cover with parchment paper and an overturned bowl or plastic bag to keep it from drying out.

Proof the Loaf (30 minutes to 24 hours)

Retard the loaf by storing it in the refrigerator for 4 to 24 hours for the best flavor. Or let it rest at 74–80°F (23–27°C) until the challah is about 1½ times its size after shaping (30 to 90 minutes).

DAY 2 OR 3

Bake the Loaf (1 to 1½ hours)

1. Bring the challah out of the refrigerator and preheat the oven to 350°F (177°C).
2. For a shiny crust, glaze with a whisked egg. Sprinkle with poppy or sesame seeds (optional).

3. Transfer the baking sheet to a middle rack in the oven and bake 20 to 30 minutes, then check for even baking and move or rotate the baking sheet as needed. Add a second baking sheet below the challah if the bottom is browning too fast. Tent with foil or lower the rack if the top is browning too fast, paying special attention not to burn seeds if used.

4. Remove from the oven when the internal temperature is at least 195°F (90°C) and the crust is golden to chestnut brown all over, usually after 35 to 50 minutes total baking time.

5. Cure the crust in the cooling oven for 5 to 10 minutes if desired (skip this step if you want a soft, flexible crust). Cool for an hour on a wire rack.

Deli Rye

Jewish rye was traditionally made with a sourdough starter for the characteristic gentle tartness to this flavorful and velvety bread. Rye flour makes the dough sticky, but adding additional flour will not help. Dampening your hands when handling the dough does the trick. Rye speeds up fermentation, so care must be taken not to over-proof. Deli rye flavor can be enhanced by replacing ¼ cup (60 grams) of the water with an equal amount of juice from real, fermented kosher dill pickles.

MAKES ONE LOAF

FLOUR: 450 GRAMS (1 POUND)
PRE-FERMENTED FLOUR: 17%
HYDRATION: 66%
WHOLE GRAIN: 33%
DOUGH: 757 GRAMS (1.7 POUNDS)

DAY 1

Build the Levain (10 minutes)

1. In a small bowl or pint jar, mix together:

10 grams starter (2 teaspoons; no need to have fed it recently for this recipe)

70 grams whole rye flour (⅜ cup)

70 grams water (¼ cup plus 1 tablespoon)

2. Cover and let it rest at 80–85°F (27–29°C) for 8 to14 hours until it has passed its peak and started to deflate. (It should still pass the float test.)

DAY 2

Mix the Dough (10 minutes)

1. Mix the flours in a large heavy bowl:

295 grams bread flour (2½ cups)

80 grams rye flour (⅝ cup)

2. Whisk together in the levain bowl or jar:

223 grams water or pickle juice/water (1 scant cup) at 90°F (32°C)

150 grams levain (all)

3. Mix the wet ingredients into the flour, combining completely until all the flour is wet, but do not knead. The dough should seem very sticky and somewhat stiff.

Autolyse (30 to 60 minutes)

1. Rest the dough for 30 to 60 minutes until it has relaxed in the bowl.

2. Add:

9 grams salt (1½ teaspoons)

16 grams caraway seeds (2 tablespoons; optional)

continued

3. Sprinkle the seeds over the dough and mix them in by poking and cutting them into the dough with a wet spatula, then folding the edges of the dough over the center with a spatula or wet hands until the dough resists folding across itself, about 20 folds. Flip the ball of dough in the bowl and cover.

Bulk Fermentation (2 to 4 hours)

1. Rest the dough at 74–80°F (23–27°C), with one set of folds after the dough has relaxed from its ball shape, usually after 30 to 60 minutes. I use a wet spatula to fold since the dough is not stretchy and is very sticky. (It will not be able to form a windowpane due to the high amount of rye flour in the dough.)

2. Let it rise until the dough doubles, is bubbly, and smells tart, being very careful not to over-proof, 2 to 4 hours total time after mixing in the levain.

Shape the Loaf (15 minutes)

1. Turn the dough out onto a lightly misted counter and pre-shape: fold the dough into thirds, like a letter, fold in the other two sides, and flip the ball. Rotate it to tighten the skin. Let it rest 10 to 15 minutes until it relaxes. The dough may seem somewhat shaggy.

2. Shape into a batard, being careful not to tear the skin. There will not be a tight skin on this loaf due to the high amount of rye flour. Flour the top generously. Cornmeal can be used if desired.

3. Place the dough seam side up in a well-floured proofing basket. Sprinkle it with flour, especially around the edges. Lay a sheet of parchment paper on top.

Proof the Loaf (30 to 60 minutes)

Cover and let the loaf rest for 30 to 60 minutes until it is risen and a little jiggly like a water balloon, but be careful not to over-proof. Move it to the refrigerator if it is ready before your oven is hot.

Bake the Loaf

1. Preheat the oven to 475°F (246°C) 30 to 60 minutes before baking, and prepare your baking and steam setup.

2. Turn out the loaf and mist it generously. Sprinkle the top with caraway seeds, cornmeal, or a pinch of sea salt if desired and score with three or four diagonal cuts across the loaf. Transfer to the oven using your chosen baking and steam setup.

If baking with a cover: Cover and bake 20 minutes for the steam phase, then uncover and continue baking until the interior is 207°F (97°C) and the crust browns to your liking, usually 35 to 50 minutes total baking time.

If baking with a steam pan: Place the loaf in the oven, then pour 1½ cups (355 milliliters) of water into the steam pan. Bake 20 minutes without opening the oven for the steam phase, then continue baking until the interior is 207°F (97°C) and the crust browns to your liking, usually 35 to 50 minutes total baking time.

3. Cure in the cooling oven with the door ajar for 5 to 10 minutes. Cool 1 to 2 hours on a rack.

KITCHENS

CHAPTER 12

Rich Dough for Something Sweet

Once you are in possession of a lively, active sourdough starter and have learned to use it confidently, it can be hard to bring yourself to use commercial yeast, even in recipes that you don't want to taste at all sour. In this chapter, I show you how to make Rich Fluffy Dough. It's a beautiful enriched dough—leavened with your sourdough starter—that's full of butter, milk, eggs, and honey. This dough is ideal for soft, fluffy breads and rolls.

You can use this dough to make all the recipes in this chapter, from a simple classic Brioche to Apple Monkey Bread and Cinnamon Rolls with Orange Icing. But don't stop there! Beyond the recipes in this chapter, any recipe you find that falls in the family of viennoiserie pastry, from babka and donuts to kulich and Pane di Pasqua, you can make using Rich Fluffy Dough. If you are an advanced or ambitious pastry baker, you can even laminate this dough with butter to make croissants and Danishes.

This dough responds very well to a cold retard, even a very long one. Combined with the fact that a double batch fits nicely in a stand mixer, this happy characteristic allows you to mix up a big batch of dough and to bake a treat, while reserving half of the dough to bake a different treat a day or two later.

Rich Fluffy Dough

I find it best to make this recipe using a stand mixer. The machine-kneaded dough is easier to mix and handle, and it makes better bread. The enrichments in the dough make it very sticky until the gluten is well developed. The enrichments also inhibit the gluten network from forming, so kneading is required in this recipe. The recipe can be doubled and still fit in a stand mixer. To minimize sourness, use freshly fed and risen levain, and use optimal yeast temperatures for the bulk fermentation. For even better results, refresh your starter the day before beginning the recipe so yeast growth is optimal. The butter is added after the autolyse because, like salt, it inhibits formation of the gluten network that's required for a nice stretchy crumb. Retarding the dough develops more flavor.

Follow this recipe through the bulk rise step, then use the dough to make one of the other recipes in this chapter.

MAKES 826 GRAMS (1.8 POUNDS) OF DOUGH

TOTAL FLOUR: 450 GRAMS (1 POUND)

PRE-FERMENTED FLOUR: 17%

HYDRATION: 60%

WHOLE GRAIN: 0%

DOUGH: 826 GRAMS (1.8 POUNDS)

DAY 1

Build the Levain (10 minutes)

1. In a small bowl or pint jar, mix together:

30 grams starter that floats and was recently refreshed (2 tablespoons)

60 grams water (¼ cup)

60 grams all-purpose flour (½ cup)

2. Cover and let it rest at 60–70°F (16–21°C) for 6 to 8 hours until it is just risen and bubbly and can float in water. Put out the butter and eggs so they will be room temperature when you mix the dough.

DAY 2

Mix the Dough (15 minutes)

1. Measure in a large heavy bowl or in the bowl of your stand mixer and set aside:

375 grams all-purpose flour (3⅛ cups)

2. Whisk together in a medium bowl until completely uniform:

2 large eggs, whisked (100 grams, ½ cup)

95 grams milk, scalded and cooled to 90°F (32°C), or water (⅝ cup)

40 grams honey (2 tablespoons)

150 grams ready levain (all, ⅔ cup)

3. Add the egg mixture to the flour and combine completely with a silicone spatula or a mixer until all the flour is wet and it forms a ball. This usually takes 2 to 3 minutes. The dough will feel very sticky and soft. Do not add additional flour or water.

Autolyse (30 minutes)

1. Cover the bowl with a lid or plate and let the dough rest for 30 minutes, until it is relaxed and softened.
2. Add:

    57 grams softened unsalted butter (¼ cup)

    9 grams salt (1½ teaspoons)

3. Spread the butter on top of the dough and sprinkle on the salt. In the mixer: knead on low for 5 to 10 minutes. It will not really form a ball but will remain stuck to the bottom. By hand: work the salt and butter through the dough completely in the bowl using a wet silicone spatula and folding the dough into the center from the sides. Fold the dough in the bowl about 100 times, using an oiled hand to finish if necessary. However you mix it, the dough should seem uniformly mixed and still very soft but a bit less sticky.
4. Form the dough into a ball with oiled hands or a bench knife and place in an oiled bowl. Cover with a lid or plate.

Bulk Fermentation (3 to 24 hours)

1. Rest the dough at 74–80°F (23–27°C). After 30 to 60 minutes, when the dough has relaxed, perform a set of stretch and folds, then flip the ball. Your dough should feel like it is getting stronger and somewhat manageable. For a stronger, stretchier dough, repeat this rest and stretch and fold another one or two times.
2. Refrigerate the dough for 8 to 24 hours. Or continue letting it rest until the dough is about double or triple its original volume, smells more like bread than flour, and a finger poke fills back in slowly (3 to 5 hours total time after mixing in the levain).

Begin Your Chosen Recipe

At this point, the dough is ready to use in one of the recipes in this chapter.

Note: If you prefer a more open crumb, preserve the loft in the dough as you shape the dough in the recipe. If you prefer a more even, closer crumb, perform some folds at the end of the bulk formation to de-gas the dough before shaping it in the recipe and be sure to allow it sufficient time to rise during the proofing step.

Brioche

This is a wonderful bread for a fancy breakfast. The crust has a tender, crispy bite and the crumb is soft and rich with buttery flavor. It's a perfect complement to your café au lait. Leftovers make incredibly good French toast, grilled ham and cheese sandwiches, bread pudding, or stuffing.

For a more traditional French presentation, the dough can be formed into traditional brioche shapes, brushed with an egg wash, and sprinkled with pearl sugar before baking.

MAKES ONE LOAF

1 prepared Rich Fluffy Dough just after bulk fermentation step (page 244; see Note on page 245)

Shape the Loaf (10 minutes)

For a plain loaf: Turn the dough out onto a lightly floured counter, shape into a batard, and gently tighten the skin, being careful not to tear the skin. Place the loaf seam side down in a greased loaf pan.

For a traditional loaf: Turn the dough out onto a lightly floured counter and divide it into eight pieces weighing about 100 grams (3½ ounces) each. Form each piece into a ball and tighten its skin. Nestle the balls in the bottom of a greased loaf pan, lining up four along each of the long sides of the pan.

Proof the Loaf (30 minutes to 24 hours)

Dust the dough with flour, cover lightly, and let it rest at room temperature for 30 to 90 minutes until it has risen to the top of the pan and it passes the poke test. Or, if the dough was not already retarded, refrigerate for 8 to 24 hours covered with parchment paper and a puffed-up plastic bag.

Bake the Loaf (45 to 60 minutes)

1. Preheat the oven to 350°F (177°C) and bring the brioche out to warm up if retarded.
2. For a shiny crust, glaze with a whisked egg. Sprinkle with pearled or sanding sugar, if desired. Score the top of the loaf with a single line (optional).
3. Bake for 45 to 60 minutes, checking halfway through for even baking and adjusting as needed. If the bottom browns too quickly, place a second pan underneath, and if the top browns too quickly, tent with foil. Remove when the interior is at least 195°F (91°C) and the crust is a honey brown all over.
4. Cool 1 to 2 hours on a rack.

Makowiec (Polish Poppy Seed Roll)

Poppy seed rolls are a vibrant part of the holiday food tradition of numerous cultures of Eastern Europe. There are many subtle variations and this is one version. While this one may not seem traditional to some, it certainly delivers in taste. Undoubtably, Makowiec was made with wild yeast before the advent of commercial yeast, so perhaps this recipe is traditional in that sense.

Some recipes call for ground almonds, orange or lemon zest, vanilla, or raisins to be added to the poppy seeds, and you can certainly add these if desired. This bread isn't overly sweet, but it is addictively delicious. For more sweetness you could add a powdered sugar glaze.

MAKES 2 ROLLS

1¾ cups poppy seeds (250 grams)

¼ cup boiling water (60 grams)

½ cup sugar (100 grams)

¼ cup soft butter (57 grams)

2 large eggs (100 grams)

1 prepared Rich Fluffy Dough just after bulk fermentation step (page 244; see Note on page 245), chilled

Prepare the Filling (45 minutes)

1. Grind the poppy seeds in a blender or spice grinder. Pour the ground poppy seeds into a small bowl and stir in the boiling water. Let sit for 30 minutes to soften.

2. Add the sugar, butter, and eggs to the poppy seed mixture and whip until thick and frothy.

Form the Rolls (45 minutes)

1. Turn out the chilled dough onto a well-floured counter and stretch into a rough rectangle. Sprinkle flour on the top and divide into two pieces. Roll out each piece into a 9-by-16-inch (22-by-40-centimeter) rectangle, using flour as needed to keep dough from sticking. If the dough keeps shrinking, let it rest 10 minutes then roll again.

2. Spread half of the poppy seed filling on each rectangle, leaving a 1-inch (3-centimeter) clean border around three sides and a 2-inch (5-centimeter) border on the 9-inch side closest to you.

3. With floured hands, roll the dough toward the 2-inch border and pinch the seam closed tightly. Roll the dough so that the seam side is on the bottom. Pinch the dough closed on each side and tuck the ends under the roll. Let the dough rest for a minute or two to seal the seams so they don't open during baking.

4. Transfer the rolls onto a parchment paper–lined baking sheet and let them rest for 20 minutes.

Bake the Rolls (1½ hours)

1. Preheat the oven to 350°F (177°C).

2. Score the top of the rolls with a single line (optional, but this can help keep the rolls from bursting if not sealed well).

3. Bake for 55 to 65 minutes, checking halfway through for even baking and adjusting as needed. If the bottom browns too quickly, place a second pan underneath, and if the top browns too quickly, tent with foil. Remove when the interior is at least 195°F (91°C) and the crust is a honey brown all over.

Hazelnut Buns

The heady combination of chocolate, orange, apricot, and roasted hazelnuts make these buns irresistible. They taste like a treat, but they will stick to your ribs. Try them with your morning coffee or take them along on a hike—they are way better than an energy bar!

MAKES 8 BUNS

1 prepared Rich Fluffy Dough just after the salt and butter are fully mixed in (page 244)

1 cup roughly chopped hazelnuts, dry roasted and unsalted (130 grams)

½ cup chopped 60% dark chocolate or chocolate chips (100 grams)

½ cup chopped dried apricots (80 grams)

zest of one orange

Complete the Dough and Bulk Fermentation (9 to 16 hours)

1. Sprinkle the hazelnuts, chocolate, apricots, and orange zest onto the dough. Fold the dough with wet or oiled hands or a spatula for several minutes until the ingredients are evenly dispersed throughout the dough, or mix briefly in a stand mixer. Form the dough into a ball with wet hands or a bench knife and place in an oiled bowl. Cover with a lid or plate.
2. Rest the dough at 74–80°F (23–27°C). After 30 to 60 minutes, when the dough has relaxed, perform a set of stretch and folds, then flip the ball. Your dough should feel like it is getting stronger and somewhat manageable. For a stronger dough, repeat this rest and stretch-and-fold process.
3. Refrigerate the dough for 8 to 16 hours.

Shape and Proof the Buns (1 to 1½ hours)

1. Turn the dough out onto a floured counter and divide in half with a bench knife, then divide each half into four for a total of eight pieces.
2. With floured hands, lightly shape each piece into a ball, being careful to tuck any chocolate pieces into the dough so they don't scorch in the oven.
3. Place the balls seam side down on a parchment paper–lined baking sheet with room to rise, sprinkle with flour, and cover to prevent drying.
4. Rest the buns at room temperature for 30 to 90 minutes until puffed up and jiggly.

Bake the Buns (1 hour)

1. Preheat the oven to 350°F (177°C).
2. Place the baking sheet with the buns in the hot oven.
3. Bake for 20 minutes, then check the buns to see if the crust is becoming too brown and adjust as needed.
4. Continue baking for a total of 30 to 45 minutes until at least 195°F (91°C) inside.
5. Cool briefly and serve warm.

Nutmeg Puffs

Any kind of fried dough is an irresistible treat, and this simple little donut is no slacker. These puffs are divine when enjoyed while still slightly warm.

Hot oil requires a careful touch to prevent burning the chef, and it should be monitored with a thermometer to be sure it is at the ideal temperature. Fine sugar, which sticks best to the puffs, can be purchased, or it can be made from regular granulated sugar by running it in the blender for a few minutes.

MAKES ABOUT 40 DONUTS

1 prepared Rich Fluffy Dough just after bulk fermentation step (page 244), chilled

½ cup fine sugar (100 grams)

1 teaspoon nutmeg, freshly ground (2 grams)

1 quart refined vegetable oil, such as sunflower (1 liter)

Shape and Proof the Donuts (1 to 2 hours)

1. Using a bench knife, divide the dough into 20-gram (¾-ounce) pieces. You should end up with about 40 or so.

2. Roll each piece into a ball and arrange the balls on an oiled or parchment paper–lined tray.

3. Cover the pieces with parchment paper and a lightly damp towel or plastic bag to prevent drying and let them rest at 75–80°F (24–27°C) for 30 to 90 minutes until the donuts have risen and your finger imprint fills in slowly if you poke the dough.

4. Combine the sugar and nutmeg in a small bowl and set aside.

Fry the Donuts (1 hour)

1. Heat the oil in a heavy 2-quart kettle over medium-high heat until the temperature reaches 375°F (191°C) and adjust heat to maintain this temperature.

2. Gently drop four to six balls into the hot oil, making sure not to crowd them (which would bring the oil temperature down). Fry until golden brown on each side, about 3 minutes total time. Pull the donuts out of the oil with a strainer and let them cool on a wire rack until they are all fried. Turn off the heat under the oil.

3. As soon as the donuts are cool enough to handle, roll each one into the nutmeg-sugar mixture and return them to the rack to finish cooling.

Apple Monkey Bread

Apples and nuts are layered between delicate puffs of brioche encased in a light caramel coat in this pull-apart bread. Perfect for a special breakfast or brunch, imagine tossing this in the oven and enjoying the aroma as it bakes on a holiday morning. You can assemble it the day before and let it retard overnight—then when you're ready to bake, just preheat the oven and pop it in. This bread can be elegantly sliced, but it's designed to be pulled apart with the fingers, puff by puff, making it a favorite of the young (and young at heart).

1 prepared Rich Fluffy Dough just after bulk fermentation step (page 244), chilled

¾ cup chopped walnuts (80 grams)

2½ cups tart apples, diced (300 grams)

1 teaspoon ginger (2 grams)

½ cup melted butter (114 grams)

¾ cup sugar (150 grams)

1 teaspoon cinnamon (2 grams)

Form the Loaf (20 minutes)

1. Turn out the dough onto a lightly floured counter and pre-shape it. Allow it to rest 5 minutes.

2. Prepare a loaf pan by greasing it or lining with parchment paper. Mix the walnuts, apples, and ginger together in a bowl and set aside. Put the melted butter into a small bowl, and mix the sugar and cinnamon in another bowl.

3. Using the bench knife, divide the dough in half, then in half again, continuing until you have about 30 pieces.

4. One by one, dip the dough pieces into the butter, then roll in the cinnamon-sugar mixture, then arrange them in the bottom of the loaf pan. When you have about 10 arranged, sprinkle half of the apple mixture into the pan. Continue with the next 10 dough pieces, then the rest of the apples, then the rest of the dough pieces. Drizzle any extra butter or sugar over the loaf.

Proof the Loaf (30 minutes to 12 hours)

Cover the loaf lightly and let it rest at room temperature for 30 to 90 minutes until risen and puffy. Or, if the dough was not already retarded, refrigerate 4 to 12 hours, but bring the Monkey Bread to room temperature an hour or so before baking.

Bake the Loaf (1½ hours)

1. Preheat the oven to 350°F (177°C).

2. Bake for 55 to 65 minutes, checking halfway through for even baking and adjusting as needed. If the bottom browns too quickly, place a baking sheet underneath, and if the top browns too quickly, tent with foil. Remove when the interior is at least 195°F (91°C) and the crust is a honey brown all over.

3. Cool on a rack for an hour.

Cinnamon Rolls with Orange Icing

This recipe makes cinnamon rolls with a little something extra. The addition of fragrant orange zest and crunchy almonds truly heightens the cinnamon experience. Excellent as brunch fare or as afternoon snacks, add more or less icing to suit your preference.

These cinnamon-forward rolls are designed to be subtly rich and sweet. For a more decadent treat, use the larger amounts of sugar and butter to stuff the rolls. If you retarded the dough, let it warm up a bit before you try to stretch it.

MAKES 13 ROLLS

1 recipe Rich Fluffy Dough just after bulk fermentation step (page 244), at room temperature

¼ to ½ cup soft butter (28 to 57 grams)

zest from one orange

¼ to ½ cup sugar (52 to 104 grams)

2 tablespoons cinnamon (12 grams)

½ teaspoon vanilla extract (3 milliliters)

1 tablespoon orange juice (15 milliliters)

½ cup powdered sugar (60 grams)

¼ cup toasted almonds, chopped or slivered (30 grams)

Form the Rolls (1½ hours)

1. Turn out the dough onto a lightly floured counter and fold in two of the sides, like a letter. Gently stretch the dough into a 13-by15-inch (33-by-38-centimeter) rectangle, using flour as needed to keep the dough from sticking to the counter. Aim for even thickness throughout the rectangle. If the dough keeps shrinking, let it rest 10 minutes, then stretch again. Alternatively, you could use a rolling pin.

2. Combine the softened butter, orange zest, sugar, and cinnamon in a small bowl. Spread it onto the dough evenly and gently, leaving a clean 2-inch (5-centimeter) border on the short side closest to you (an offset spatula makes this easier).

3. With floured hands, roll the dough toward the clean border and pinch the seam closed tightly along its length. Wrap the roll in parchment paper and chill in the refrigerator for an hour to make the dough easier to cut cleanly (optional).

4. Cut the roll into 13 even slices (you can mark them in advance with a shallow cut using a ruler). Nestle them in an 8-by-13-inch (20-by-33-centimeter) baking pan, buttered or lined with parchment paper.

Proof the Rolls (30 minutes to 24 hours)

Cover the rolls lightly and let them rest at room temperature for 30 to 90 minutes until risen and puffy. Or, if the dough was not already retarded, refrigerate 4 to 24 hours.

Bake the Rolls (1 hour)

1. Preheat the oven to 375°F (177°C).

2. Bake for 20 minutes, then check the rolls to see if the crust is becoming too brown and adjust as needed. If the bottom browns too quickly, place a second pan underneath, and if the top browns too quickly, tent with foil.

3. Continue baking for a total of 25 to 35 minutes until at least 195°F (91°C) inside and the crust is honey brown.

4. Cool until just warm.

Ice the Rolls

Mix the vanilla and orange juice into the powdered sugar to make an icing and spread or drizzle it over the warm rolls. Immediately sprinkle with the almonds before the icing sets.

Chocolate Raspberry Buns

Bittersweet chocolate and tart, fragrant raspberries melt together in the pillowy softness of these buns. Perfect for an extra-special breakfast or afternoon tea (perhaps on Valentine's Day), you can roll them up and store them in the refrigerator, then slice and bake them the next day.

MAKES EIGHT BUNS

1 recipe Rich Fluffy Dough just after bulk fermentation step (page 244), at room temperature

¾ cup chocolate chips (120 grams)

1 cup fresh or frozen, unthawed raspberries (115 grams)

1 egg, whisked (optional)

sanding sugar (optional)

Form the Buns

1. Turn out the dough onto a lightly floured counter and fold in two of the sides, like a letter. Gently stretch the dough into a 9-by-14-inch (23-by-36-centimeter) rectangle, using flour as needed to keep the dough from sticking. If the dough keeps shrinking, let it rest 10 minutes, then stretch again. Alternatively, you could use a rolling pin.

2. Distribute the chocolate and raspberries on the dough evenly, leaving a clean 2-inch (5-centimeter) border on the short side closest to you.

3. With floured hands, roll the dough toward the 2-inch border and pinch the seam closed tightly. Aim to minimize air gaps by pressing the dough into the filling as you roll. Wrap the roll in parchment paper and chill in the refrigerator for 1 hour to make the dough easier to cut cleanly (optional; chilling is not recommended if frozen berries are used).

4. Cut the roll into eight even slices and arrange them on a parchment paper–lined baking sheet with plenty of space between them for expanding.

Proof the Buns (30 minutes to 24 hours)

Cover the buns lightly and let them rest at room temperature for 30 to 90 minutes (2 to 3 hours if frozen berries are used) until risen and puffy. Or, if the dough was not already retarded, refrigerate 4 to 24 hours.

Bake the Buns (1 hour)

1. Preheat the oven to 375°F (191°C).
2. For a shiny crust, glaze with a whisked egg. For sparkle, sprinkle with sanding sugar.
3. Place the baking sheet with buns in the hot oven.
4. Bake for 20 minutes, then check the buns to see if the crust is becoming too brown and adjust as needed. If the bottom browns too quickly, place a second pan underneath, and if the top browns too quickly, tent with foil.
5. Continue baking for a total of 30 to 45 minutes until at least 195°F (91°C) inside and the crust is a pale honey brown.
6. Cool on a rack until just warm.

Part Five
Appendix

Waste Not, Want Not: Recipes Using Discarded Starter

Once you have established a routine of baking and maintaining your starter, you likely won't have much sourdough discard. If you are finding you have a lot of starter discard on a regular basis, you may want to try a different starter management routine or to keep a smaller amount of starter.

But when you do have a mess of discard, what can you do with it? As the next few pages demonstrate, you can use it as the base for a new recipe. We will make tantalizingly delicious waffles, pancakes, and crackers that use up a lot of discard. Alternatively, you can add small amounts of discard to add a bit of flavor to any recipe that includes flour, such as dumplings, cakes, biscuits, deep-fry batters, and quick breads. Just mix it in with the wet ingredients.

It is not advisable to use discard from the first three or four of the feedings of a brand-new starter because the necessary populations of beneficial bacteria are not yet present to prevent the growth of potentially harmful organisms. Wait until the new starter is actively bubbly and smells pleasantly acidic before consuming its discard.

Wild Waffles

The sourdough discard that is the base of these quick, easy waffles is full of wild yeast and bacteria, but what's really wild is what happens when you add baking soda. Depending on how acidic your discard is, a little pinch of baking soda might be plenty. If you add too much baking soda, there might be a small volcano in your kitchen!

The addition of a small amount of spelt or buckwheat flour adds a flavor boost, but you can substitute any other flour you like.

MAKES ABOUT 1½ CUPS (350 MILLILITERS) OF BATTER, ENOUGH FOR 2 TO 4 WAFFLES, DEPENDING ON YOUR WAFFLE IRON. THIS RECIPE ASSUMES YOUR DISCARD IS AT 100% HYDRATION.

TOTAL FLOUR: 130 GRAMS (0.3 POUND)

PRE-FERMENTED FLOUR: 88%

HYDRATION: 165%

WHOLE GRAIN: 12%

DOUGH: 356 GRAMS (0.8 POUND)

AHEAD OF TIME

Collect Sourdough Discard (up to 2 weeks)

Store the discard from established starter refreshments in a loosely covered pint jar in the refrigerator until you have enough. Don't worry if it forms colored hooch or continues bubbling.

WAFFLE DAY

Mix the Batter (5 minutes)

1. Preheat and prepare your waffle iron according to its instructions.

2. Whisk vigorously with a fork in the jar until uniform consistency:

230 grams discard (1 cup)

3. Add to the discard jar and continue whisking until smooth:

100 grams eggs (2 large)

15 grams spelt or buckwheat flour (2 tablespoons)

2 grams salt (¼ teaspoon)

8 grams honey or sugar (1 teaspoon; optional, promotes browning)

The batter should be thick but pourable, and the consistency will vary due to differences in the starter discard consistency. If the batter is runny, continue adding flour, a spoonful at a time, and whisking it in until it thickens. If the batter seems too thick, add water or milk.

4. Add, a pinch at a time, and whisk vigorously until the batter begins to bubble:

> **1 gram or less baking soda (⅛ teaspoon)**

Cook the Waffles (10 minutes)

Cook the batter in your waffle iron according to its instructions.

Forty-Niner Flapjacks

Sourdough pancakes have a heartier, chewier texture than typical quick-bread pancakes, and they pack a lively, complex flavor that goes very well with a hot berry compote. In 1849, the gold seekers who came to the American West often carried sourdough in a leather pouch around their neck to keep it safe and warm as they traveled. They would mix up their dough and sleep with it during cold weather to be assured of sufficient fermentation for their morning meal.

Lucky for us, these pancakes whip right up on the fly with our pre-fermented discard. You can even mix it the night before, like a forty-niner, for a quicker breakfast in the morning. This recipe assumes your discard is at 100% hydration.

MAKES 12 TO 15 PANCAKES

TOTAL FLOUR: 148 GRAMS (0.3 POUND)

PRE-FERMENTED FLOUR: 39%

HYDRATION: 151%

WHOLE GRAIN: 0 TO 61%

DOUGH: 382 GRAMS (0.8 POUND)

AHEAD OF TIME

Collect Sourdough Discard (up to 2 weeks)

Store the discard from established starter refreshments in a loosely covered pint jar in the refrigerator until you have enough. Don't worry if it forms colored hooch or continues bubbling.

FLAPJACK DAY

Mix the Batter (5 minutes)

1. Whisk vigorously with a fork in the discard jar until uniform consistency:

115 grams discard (½ cup)

115 grams milk, vegan milk, buttermilk, or water (½ cup)

50 grams egg (1 large)

2. Add to the jar and continue whisking until smooth:

90 grams flour (¾ cup; buckwheat, whole wheat, corn, spelt, white)

2 grams salt (¼ teaspoon)

8 grams honey or sugar (1 teaspoon; optional, promotes browning)

The batter should be thick but pourable, and the consistency will vary due to differences in the starter discard consistency and the type of flour and milk chosen. If the batter is too runny, add flour a spoonful at a time, whisking it in until it thickens. If the batter is too thick, add water or milk a spoonful at a time, whisking it until it is smooth. (Batter can be held at ambient temperature or refrigerated overnight at this point, but it may rise, so transfer it to a larger container. Baking soda is optional if your batter is risen.)

3. Add and whisk vigorously until the batter begins to bubble:

2 grams baking soda (¼ teaspoon)

Cook the Flapjacks (10 minutes)

Fry the batter in a lightly oiled and preheated frying pan over medium-high heat. Flip the pancakes when the batter is dull around the edges and the bubbles have popped.

VARIATIONS

COCONUT–CHOCOLATE CHIP: Add ¼ cup (18 grams) unsweetened shredded coconut to the batter with the flour. Just after pouring the batter into the pan, nestle 4 or 5 chocolate chips in each pancake before flipping. Nice with whipped cream or even plain.

BANANA OR BLUEBERRY: Just after pouring the batter into the pan, nestle 4 or 5 berries or banana slices in each pancake before flipping. These classics beg for real maple syrup.

ALMOND: Add ¼ cup (30 grams) ground almonds and ¼ teaspoon (1 gram) almond extract to the batter with the flour. Pairs nicely with fresh plum or cherry compote.

PECAN: Add ¼ cup (30 grams) toasted, chopped pecans to the batter with the baking soda. Excellent with fresh peach slices, maple syrup, and bacon.

STRAWBERRY ROSE: Add 1 teaspoon (5 grams) rose water to the batter with the wet ingredients. Just after pouring the batter into the pan, nestle 4 or 5 strawberry slices in each pancake before flipping. Delicious with honey yogurt or whipped cream.

Parmesan Lemon Crackers

These snappy little crackers are great to eat on their own or complemented by a dip. Tuscan white bean, spinach, or artichoke dips are all fine choices. The whole grain gives these crackers a crispy crunch, and the lemon-Parmesan combination is nearly addictive.

Go for a highly aged Parmesan cheese if you can find it. The flavor punch will be worth it. Your favorite weak-gluten wheat, such as einkorn, emmer, durum, or spelt, will perform beautifully in these crackers because gluten development is undesirable in a cracker. I like to use durum. This recipe assumes your discard is at 100% hydration and is made with all-purpose flour.

MAKES 1 QUART (1 LITER) OF CRACKERS

TOTAL FLOUR: 250 GRAMS (0.6 POUND)
PRE-FERMENTED FLOUR: 46%
HYDRATION: 48%
WHOLE GRAIN: 54%
DOUGH: 405 GRAMS (0.9 POUND)

AHEAD OF TIME

Collect Sourdough Discard (up to 2 weeks)

Store the discard from established starter refreshments in a loosely covered pint jar in the refrigerator until you have enough. Don't worry if it forms colored hooch or continues bubbling.

CRACKER DAY

Mix the Dough (5 minutes)

1. Whisk vigorously with a fork until uniform consistency:

230 grams discard (1 cup)

2. Combine in a medium bowl:

135 grams whole wheat flour (1⅛ cups; any kind of wheat)

zest of 1 lemon, grated

5 grams fresh lemon juice (1 teaspoon)

30 grams freshly grated Parmesan cheese (½ cup)

freshly ground black pepper to taste

5 grams salt (¾ teaspoon)

3. Mix in the discard until well combined and fold the dough in the bowl several times. This usually takes around 3 minutes, and the dough should feel smooth and firm, like pie dough. Rest the dough for 20 minutes or so.

Roll the Crackers (10 minutes)

1. Turn out half of the dough onto a floured sheet of parchment paper. Pat the dough into a rectangle, sprinkle it with flour, and roll it out with a rolling pin to ⅛ inch (3 millimeters) thick. Transfer the parchment paper and dough to a baking sheet and cut the dough into crackers with a pizza cutter or bench knife.
2. Repeat with the second half of the dough.

Bake the Crackers (20 to 30 minutes)

1. Fifteen minutes before baking, preheat the oven to 350°F (177°C) and arrange two racks in the centermost slots.
2. Bake the crackers 25 to 40 minutes, switching the position of the pans halfway through. Toward the end of baking, watch for the crackers on the edges of the baking sheet to finish baking earlier, and remove them so they don't burn. Crackers are done when lightly golden brown. They may seem soft but will crisp up as they cool.
3. Cool them completely on a wire rack. Store in an airtight jar.

Sesame Thyme Crackers

Excellent eaten out of hand or paired with fresh vegetables, slow-roasted tomatoes, hummus, or baba ghanoush. The soul-satisfying base of the sesame is accented by the bright combination of sumac and thyme.

Tahini is simply sesame seeds ground into a paste. It can be found, along with sumac, at Middle Eastern or specialty stores. If you can't find tahini, substitute natural almond butter. If you can't find sumac, substitute lemon juice. I make my crackers with whole durum flour, but you can use any kind of whole wheat flour. This recipe assumes your discard is at 100 percent hydration and is made with all-purpose flour.

MAKES 1 QUART (1 LITER) OF CRACKERS

TOTAL FLOUR: 250 GRAMS (0.6 POUND)

PRE-FERMENTED FLOUR: 46%

HYDRATION: 46%

WHOLE GRAIN: 54%

DOUGH: 422 GRAMS (0.9 POUND)

AHEAD OF TIME

Collect Sourdough Discard (up to 2 weeks)

Store discard from established starter refreshments in a loosely covered pint jar in the refrigerator until you have enough. Don't worry if it forms colored hooch or continues bubbling.

CRACKER DAY

Mix the Dough (5 minutes)

1. Whisk together vigorously with a fork until uniform consistency:

230 grams discard (1 cup)

40 grams sesame tahini (2 tablespoons)

2. Combine in a medium bowl:

135 grams whole wheat flour (1⅛ cups; any kind of wheat)

8 grams sumac (2 teaspoons)

2 grams dried thyme (2 teaspoons)

7 grams salt (1 teaspoon)

freshly ground black pepper to taste

3. Mix in the discard mixture just until well combined and fold the dough in the bowl several times. The dough should feel smooth. Rest the bowl in the refrigerator for 20 minutes or so.

Roll the Crackers (10 minutes)

1. Turn out half the dough onto a floured sheet of parchment paper. Pat the dough into a rectangle, sprinkle evenly with 12 grams (1 tablespoon) sesame seeds, and roll it out out with a rolling pin to ⅛ inch (3 millimeters) thick. Transfer the parchment paper and dough to a baking sheet and cut the dough into crackers with a pizza cutter or bench knife. Optional: sprinkle the crackers with a few shakes of additional salt.
2. Repeat with the second half of the dough.

Bake the Crackers (20 to 30 minutes)

1. Fifteen minutes before baking, preheat the oven to 350°F (177°C) and arrange two racks in the centermost slots.
2. Bake the crackers 25 to 40 minutes, switching the position of the pans halfway through. Toward the end of baking, watch for the crackers on the edges of the baking sheet to finish baking earlier, and remove them so they don't burn. Crackers are done when lightly golden brown. They may seem soft but will crisp up as they cool.
3. Cool them completely on a wire rack. Store in an airtight jar.

Special Baking Challenges

What should you do if you want to bake during a heat wave, or you're at 7,000 feet above sea level? Here are a few tips for some common baking challenges.

Bread Making in Sweltering Heat

For successful fermentation, put your dough in a cooler next to a bowl of ice and use a thermometer to check the temperature of the dough to make sure that it stays in the range called for in the recipe. Or make use of a cellar, cool garage, or a room on the north side of the building. You could also immerse the bottom of the bowl in a sink or tub of cool water and add ice as needed. Be careful not to get water in the dough.

For baking without heating up the kitchen, nothing beats a grill (except a wood-fired pizza oven, which should come with instructions for bread baking). The two main issues with baking on a grill (which I will help you avoid) are scorching the bottom and unwanted smoky taste. Note: The grill must have a cover, which you will keep closed during preheating and baking.

- Use a metal lidded baker, such as a Dutch oven. Place the Dutch oven on one or two baking sheets to prevent the bottom of the loaf from browning faster than the rest of the crust. Put the lid on the Dutch oven and close the grill. Bake the bread covered the entire time, lifting the lid of the Dutch oven after the steam phase to release the steam and then replacing the lid. This will prevent a smoky taste. Then cure the crust on the cooling open grill (as pictured).
- If your grill doesn't have a thermometer, use an oven thermometer. Use a temperature 25°F (14°C) lower than the recipe calls for. Preheat a gas grill for 15 minutes.
- Use indirect heat, such that the heat source is not directly under the loaf, but off to the sides to further prevent bottom scorching.

Bread Making at Altitude

In the mountains, air is drier (so water is lost from exposed dough), air pressure is lower (so carbon dioxide gas expands more), and water boils at a lower temperature (so bread is done baking at a lower temperature) than at sea level. This can change several things in your bread-making process:

- Flour may be drier and water may be lost from the dough, so you may need to add a little extra water to the dough. Wait until at least 30 minutes after mixing to decide if extra water is needed. Adding just 23 grams (1½ tablespoons) will raise the hydration by a whopping 5 percent, so add just a little at a time.
- Dough may rise faster than expected, which will leave the fermentation and gluten formation behind. (This is less of an issue for heavy whole grain doughs.) If you want better gluten structure for loft and a more complete fermentation for digestibility or flavor, you can do the following.
    - Lower the fermentation temperature a few degrees or employ a cold retard to slow down carbon dioxide production.
    - Continue stretch and folds later into the bulk fermentation to release some of the excess carbon dioxide gas and to develop the gluten network more quickly.
    - Use a smaller amount of starter or levain, such as 25 to 50 percent less.
- Beware of over-proofing your loaf. There will be considerable oven spring, so a loaf that is well risen going into the oven may fall. It is far better to under-proof your loaf and to let it spring in the oven.
- Take care to cover the dough at all stages to protect it from drying out and forming a crust.
- Baking may be faster by 5 to 10 minutes, so check on your bread early.
- Steaming during baking is even more important at altitude, so consider a cover for best results as covers work better than steam pans in most ovens.
- Due to the lower boiling temperature of water, your loaf will be fully cooked at a lower internal temperature than at sea level. If in doubt, try knocking on the bottom of the loaf, which will sound hollow and resonant when it's fully cooked. The expected temperature when done for artisan breads baked at altitudes above sea level are as follows:

    1,000 ft (305 m) . . . . . . 205°F (96°C)
    3,000 ft (914 m) . . . . . . 201°F (94°C)
    5,000 ft (1,524 m) . . . . . 197°F (92°C)
    7,000 ft (2,134 m) . . . . . 194°F (90°C)
    9,000 ft (2,743 m) . . . . . 190°F (88°C)
    11,000 ft (3,353m) . . . . 186°F (86°C)
    13,000 ft (3,962 m) . . . . 183°F (84°C)

- Once the bread is baked and cooled, store it in a bread box or plastic bag, or both. The cut end will become stale within minutes in dry mountain air. When using the bread, rewrap it immediately.
- Stale slices can be misted with water, then toasted just before use.

Bread Making with Freshly Milled Flour

Baking with freshly milled flour can add beautiful flavors to your bread, and it can be comforting to know exactly what's in the flour. Flavor and nutrition are at their peak right after the wheat is ground. As the flour ages, its nutritional value begins to deteriorate, and its flavor fades or becomes bitter. In this respect, grains are like most other produce—think of slicing an apple or grinding coffee beans. Unlike commercial flour, for which consistency is absolutely required for marketability, freshly milled flour can perform inconsistently in a bread recipe. Commercial mills test each batch of grain and blend them together to produce a flour with consistent baking characteristics. When grinding your own small-batch flour, performance will change depending on the type of wheat berries used and how they were grown and stored, the type of mill, the coarseness chosen, the humidity, the temperature during milling, and many other factors. You will need to keep an eye on your dough so you can adjust for inconsistencies in the flour.

If you are switching to milling your own flour, you may want to choose a recipe that works well for you and try substituting just one-quarter of the commercial flour for fresh milled. Since you are familiar with how the dough and bread should come out, you can adjust the recipe as needed. Then you can increase the proportion of freshly milled flour for your next loaf and continue your experiment until you are baking with 100 percent freshly milled flour. Some bakers routinely

use a mix of freshly milled flour with commercial flour for the best of both worlds.

Here are some of the more common challenges:

- Flour that becomes too hot during grinding can cause poor performance in bread due to starch and enzyme damage. If you are having this problem, grind chilled berries or crack berries coarsely, let them cool, then regrind to achieve a fine flour.
- Fresh flour can seem to need extra water when mixing the dough, although this can sometimes be due to slower absorption of the water. Wait until 30 minutes after mixing to adjust hydration and add water a teaspoonful at a time until it is the proper consistency for the recipe.
- Fermentation can be faster, leading to doughs that over-ferment and fall in the oven. Use a cooler temperature or a smaller amount of starter if this is happening, or retard the fermentations before the dough has fully risen and end the fermentations promptly.
- Freshly milled flour includes all the bran and germ, like a stone-ground whole wheat flour, which will make the flour

perform quite differently than white flour or even many commercial whole wheat flours. This can lead to a denser bread due to the cutting action of the bran on the gluten network and due to the higher need for water in a dough with more bran. Be sure to add enough water to get the dough consistency called for in the recipe. Follow instructions for whole grain dough handling in the Rustic Boule recipe (page 99) to prevent damage to the gluten network. Or use a 40-, 50-, or 60-mesh sieve to remove some or most of the bran and germ if you want the flour to perform more like the commercial flours you are used to.

- Is your bread not rising very high and the crust lacks flavor and color? Using diastatic malt can help if your wheat berries happen to be deficient in amylase activity. Many commercial mills routinely add diastatic malt to their flour to achieve consistent amylase activity.
- If the gluten seems weak because the dough does not become strong when you stretch and fold it and the bread has a dense crumb, adding vital wheat gluten will help. Or you can age the flour, as commercial mills do (and has been done for thousands of years). Place the flour in a paper or cloth bag for 1 to 3 weeks in a cool, dry place (or longer if you've sifted out the bran and germ, which become rancid during oxidation) to restructure the gluten through natural oxidation. You will still have flour that is much fresher than store-bought, but it will lose some of its fresh flavors.
- Freshly ground flour is fluffier, so it's best to measure by weight.

Traveling with Your Starter

When traveling to a place where you have access to an oven, you will certainly be tempted to bake some bread. Perhaps you want to present a fresh loaf to your hosts or to make sure you have something decent to eat when in unfamiliar surroundings. Never fear, you can bring your bubbly friend along, even on an airplane.

- Mix some starter with flour to form a marble-sized ball of firm dough. Bury this dough in a small container of flour, such as a tiny jar, plastic container, or a strong plastic bag. This dough will survive for many days in various temperatures without doing anything dramatic (like exploding or dying) because of the low hydration, although you should avoid freezing it or subjecting it to high temperatures (over 95°F [35°C]). For extra insurance, place the container in a plastic bag.
- When you arrive, take the ball out and break it up into pieces, stir the pieces into 30 grams (2 tablespoons) of water and stir in 30 grams (¼ cup) of flour. You can use it to bake as soon as it's risen.

Bread Making When Camping

If you are car camping, you can bring your regular starter in your cooler and bake bread in a heavy Dutch oven over a campfire or camp stove. Don't forget flour for dusting and parchment paper for easy cleanup.

If you are wilderness camping, there are several adjustments to make. It will be a bit of an adventure with a definite learning curve. It is best to use a simple slow fermentation recipe such as Slow and Easy (page 163) or 60% Whole Wheat (page 178), with the hydration reduced to 65 percent of the weight of the flour. Here's a method I like:

PREPARE IN ADVANCE:

- Preassemble the dry ingredients for a quarter recipe of the chosen bread in a 1-quart sealable plastic bag. Plan how you will measure the amount of water called for in the recipe while camping. Each bag makes a micro loaf that fits into a backpacker's pot and cooks up quickly over a lightweight stove. This loaf will feed one or two hungry hikers as an accompaniment to a meal.
- Create a 50 percent hydration firm starter that resembles a ball of stiff dough, as follows: mix 10 grams of your starter with 25 grams (3 tablespoons) flour and 10 grams (2 teaspoons) water to get

The backpacker's boule

45 grams of stiff starter. This ball, which can be placed in a small bag or container, should be hardy enough to survive many days on the trail without refreshment and will make about 6 micro loaves. Carry the starter in the center of your pack where it can be protected from excessive heat.

TO MIX THE DOUGH:

- Pour the flour mix into a pot. Tear a marble-sized piece of the firm starter into small pieces and add to the pot. Measure the water into the pot and mix well. Cover and let the dough rest for a spell, like 10 to 60 minutes, then give the dough a few folds and form a ball. Add a few drops of oil to the now empty bag to coat the interior completely and then place the ball in the oiled bag.
- Or mix the dough directly in the sealable bag by adding the starter and water to the flour, zipping the bag closed, and kneading the bag until the contents are uniform. You won't have to clean a mixing pot, but a lot of the dough might remain stuck inside the bag. Even so, don't discount this method. You can add the contents for the next day's bread to this bag and use the scraps inside as the starter just as pioneer bakers often did with their bread bowl on the trail.

FOR THE BULK FERMENTATION:

- Remove some of the air from the bag and seal it. For extra insurance, place the quart bag inside a gallon bag. I like to mix the dough the night before and hang it in the bear bag where it gets a slow, cold start, then carry it the next day, either in my pack or, if it's cold, tucked inside my jacket to keep it warm while it rises. I bake it before dinner.

TO BAKE THE DOUGH:

- For a loaf, form a boule, place it on a piece of parchment paper in a small pan, and let it rise, covered. To bake, place the small pan inside a bigger (not nonstick) pot with a few pebbles and 10 milliliters (2 teaspoons) of water in the bottom. Cover the pot and steam over a very low flame for about 15 minutes. Then release all the steam, cover, and continue dry baking over a low flame for another 5 to 15 minutes until the crumb is done. Once the crust is set and the water is boiled away, flip the loaf to cook the top crust more evenly and to prevent the bottom from scorching. The crust will not brown much except on the bottom, as pictured.
- For flatbread, tear off pieces of dough with wet hands and stretch it into circles to cook in a hot oiled or parchment paper–lined pan; flip the bread halfway through.

None of these breads will be beautiful to look at, but it won't matter since they'll taste amazing under a starry sky while you prop up your tired feet.

Troubleshooting

Dough Not Rising Fast Enough/Levain Not Ready

- Rest the dough at 74–80°F (23–27°C).
- Use a more recently fed starter at its peak.
- If you no longer can fit the baking project into your schedule because the rising time took too long, refrigerate the dough and come back to it later (up to 8 hours or possibly longer). The dough will continue to rise in the refrigerator for a couple of hours, then it will slow down dramatically as the dough chills.

Levain Doesn't Float

IF IT HASN'T RISEN SUFFICIENTLY IN THE ALLOTTED TIME:

- Is it cold? Warm it up and give it some more time.
- Was the starter you used starved or neglected? Remake the levain using this levain as the starter in a 1:1:1 proportion and let it rest at 74–80°F (23–27°C) for 2 hours.

If none of that is the problem, you may need a better starter.

IF IT ROSE, THEN DEFLATED SIGNIFICANTLY:

- Remake the levain using this levain as the starter in a 1:1:1 proportion and let it rest at 74–80°F (23–27°C) for 2 hours until it doubles.

Dough or Levain Is the Wrong Temperature

Change the temperature of your dough by placing the bowl or jar in a water bath that is just above or below the correct temperature. For example, if your dough is supposed to be at 74–80°F (23–27°C) after mixing, but when you measured the temperature it was at 68°F (20°C), you can immerse the bottom of the dough bowl in a bigger pot or bowl filled with water that is at 85°F (29°C) for 30 minutes or so, adding hot water as needed. Make sure not to go past 90°F (32°C). Conversely, dough that is too warm can be chilled with iced water using the same process. Be careful that water does not get in your dough or levain.

Dough Is Too Dry after Mixing

- Check the recipe to see if you mismeasured or forgot anything.
- Wait 30 minutes after mixing and check the dough again. Some flours, especially whole grain and durum, take a long time to seem wet.
- If it still seems dry, you may need to add water, especially if you measured with cups. Add water a teaspoon or two at a time and mix it in after each addition, adding only as much as you need to get the consistency called for in the recipe.

Dough Is Too Wet

RIGHT AFTER MIXING:

- If you are used to baking breads with commercial yeast or low hydration, keep in mind that these sourdoughs have much more water in them—the dough is supposed to be wet and sticky (but never soupy).
- Check the recipe to see if you mismeasured or forgot anything.
- You may need more flour. Add a tablespoon of flour at a time and mix it in after each addition. You don't want to add too much.
- Did you substitute a lot of spelt flour? Spelt flour cannot absorb as much water as wheat flour, so add additional flour a tablespoon at a time and mix it in after each addition.

AFTER BULK FERMENTATION:

- If the dough was fine before bulk fermentation but is soupy after, and it smells pretty sour, it has probably over-proofed. (See Dough Is Over-Proofed entry below).
- If it is soft and damp but still has structure (stays in a mound after pre-shaping), it's probably just fine.

Dough Is Too Stiff to Shape the Loaf or Is Tearing

- Let it rest for 10 minutes to allow the gluten to relax, then try again. Cold dough takes longer to relax.
- If it is still tearing and is soft, it may be over-proofed. (See the following entry, Dough Is Over-Proofed.)

Dough Is Over-Proofed

IF IT IS ONLY A LITTLE OVER-PROOFED AND STILL HAS STRUCTURE:

- Bake it very soon with support to help it rise in the oven. There is little or no benefit to scoring, and scoring may make the bread rise less in the oven.

- First, start preheating your oven.
- **Option 1:** You could proof and bake it in a loaf tin, greased or with a parchment paper liner.
- **Option 2:** To get the intended crusty boule, you could proof it in your proofing basket seam side down on a 13-inch (33-centimeter) square or larger piece of parchment paper. When your oven is hot and your dough has had 30 minutes of proofing time, transfer the parchment paper and dough into a covered baker. A smaller diameter baker will give the dough more support.

IF IT IS VERY OVER-PROOFED AND LOOSE:

- See the following entry, Rescuing a Disaster.

NEXT TIME:

- Shorten the bulk fermentation and be sure it is not above 80°F (27°C) for very long.
- Check the dough more frequently toward the end of the bulk fermentation.
- Be sure your starter is recently refreshed and review the relevant tips in Part 2 of this book to manage your starter for optimal yeast population.
- Be sure your starter can double in 2 to 3 hours at 74–80°F (23–27°C).
- Remember that whole grain dough ferments more quickly than white flour dough.

Rescuing a Disaster

No matter what's wrong, don't throw it away! You can probably rescue it or use it to make something. Peruse these three creative solutions before pitching your dough.

RESCUE OPTION 1: If you've just had it and you don't know how to fix it, and you just want this baking odyssey to be over, try this:

- Place a big piece of parchment paper into a 3- to 4-quart pot or casserole, dump all your crazy dough into it (it helps to swear at it a little, too), cover the pot, proof it if you think that will help, and place the pot in a 475°F (246°C) preheated oven.
- Remove the cover after 30 minutes of baking time and with luck there will be a nice loaf of bread in there! Continue baking until it has a nice brown crust. If it's flat, don't worry. It can still be pretty good eaten warm with some butter. Or you can slice it thinly, spread the slices on a baking sheet, and dry them in the oven at 200°F (93°C) for delicious home-baked melba toast. Or, possibly, dog biscuits.

RESCUE OPTION 2: If your dough has transformed into soup that used to be dough, try this to save it. You will end up with two possibly great, but probably just good, loaves. We will treat the dough like it's a giant levain:

- Calculate the total flour, water, and salt in the original recipe and mix that amount into your original dough (water first, then salt and flour).
- Give it a very short bulk fermentation until it is risen not quite double in size, watching it closely all the while. Then

divide it into two loaves, proof a short time, and bake. You can freeze the other loaf or give it as a gift!

RESCUE OPTION 3: If you don't want to go to the trouble, store the dough in the refrigerator and use it up by making sourdough discard recipes over the next couple of weeks. You'll get some great waffles, pancakes, crackers, and many other items you can find recipes for on the internet.

Loaf Is Stuck to the Proofing Basket

RESCUE OPTION 1: Lift the proofing basket off the loaf very slowly, little by little. It may come out after some time, but you might have to nudge it with your finger where it's stuck.

RESCUE OPTION 2: If it still won't come out, you may need to scrape it out with a spatula. The loaf may need to be reshaped and will need to proof again for a short amount of time, probably somewhere between 30 and 60 minutes.

NEXT TIME:

- Sprinkle more flour in the proofing basket and on the loaf.
- Use rice flour.
- Proof the loaf top side up on floured parchment paper and transfer directly to the baking surface without flipping it over.
- For long proofs or very wet, sticky dough, line bannetons with a smooth cotton or linen cloth, generously floured.

Bottom Crust Too Dark

- Place a baking sheet below the bread (unless it is on a stone).
- Move up to a higher rack.
- Sprinkle coarse cornmeal under the parchment paper.

Top Crust Too Dark

- Move to a lower baking rack.
- Tent a piece of foil over the top.
- Turn down the oven temperature.

Crust Not Shiny

- Provide better steam during early baking. Lack of moisture in the oven prevents the gelation of the starches on the surface of the dough.
    - Mist the loaf generously just before baking and mist the inside of your baking lid or cover.
    - Switch to using a cover instead of a steam pan, since some ovens vent the steam too rapidly. (Do not block your oven vent!)
- Aim for a more complete fermentation. Under-fermentation causes a dull crust.

Crust Not Thick Enough

- Increase the length of the steaming phase of baking.
- Bake the loaf longer, reducing the temperature if it's browning too fast.
- End bulk fermentation earlier.

Crust Too Thick

- Decrease the length of the steaming phase of baking.
- Take the loaf out of the oven sooner after it reaches the target internal temperature.
- Be sure the loaf surface is not drying out during proofing.
- If the bottom crust is too thick, try baking on a higher oven rack and baking at a lower temperature.
- Aim for a more complete bulk fermentation.

Crumb Too Open or Irregular

- Add an extra set of stretch and folds toward the end of the bulk fermentation to achieve a finer, more uniform crumb.
- Use less water in the dough.
- End bulk fermentation earlier.

Crumb Too Closed or Regular

- Be sure to leave the dough to rise undisturbed for the second half of the bulk fermentation and to retain as much loft as possible during the loaf-shaping process.
- Use more water in the dough.
- Be sure to allow a strong gluten network to form, via folds or time, and use a strong flour or add vital wheat gluten to the flour if it's weak.
- Be sure to allow large bubbles to form before ending the bulk fermentation.
- Retard the loaf.

Crumb Has Air Pocket below Crust

- This can happen when the gluten isn't strong enough in a high-hydration dough. Increase the stretch and folds early in the bulk fermentation.
- If you are still having this problem, your flour may be weak in gluten strength. Use a stronger flour, such as bread flour, or add vital wheat gluten to your flour to increase its strength. You could also decrease the hydration in the dough, but this would change the resulting bread from the recipe's intention.

Bread Not Sour Enough

- Retard the dough in the refrigerator for 12 to 24 hours either after the bulk fermentation or during the proofing of the loaves.
- If your bread is routinely under-sour for your taste, try refreshing the starter only after it peaks, when more of the bacteria have built up, or switch to feeding it only rye flour.
- Allow your starter to grow adequately before refrigerating it.

If you have done all the above and are still having too little sourness, you may want to try a different starter.

Bread Too Sour

- Perform fermentation or proofing steps at yeast's ideal temperature (74–78°F; 23–26°C) to minimize bacteria growth (the bacteria produce the acids) and be sure your starter is kept cool.
- Eliminate or shorten the time retarding the dough in the refrigerator. If acetic acid is being made, it will accumulate in the dough during long periods of refrigeration.
- If you are already using ideal temperatures and shortening (or eliminating) the retard time in the refigerator and your dough is still too sour, review your starter management.
- If you have done all the above and are still having too much sourness, you may want to try a different starter, such as the 1847 Oregon Trail Sourdough Starter, which lacks bacteria species that make a lot of acetic acid.

Poor Oven Spring

- Your dough may have over-proofed in the bulk fermentation or during loaf proofing, though loaf over-proofing is commonly the culprit. The next time, end the fermentation or loaf proofing a bit earlier, ferment at a lower temperature, or proof your loaf in the refrigerator.
- You may have forgotten to score the loaf, so the dough was unable to push out of the crust.
- The crust may have firmed up before the dough was finished rising. You may need to provide better steam during baking or to bake in an enclosed chamber such as a pot.
- Try proofing the loaf in the refrigerator and baking it while cold. This can often provide very nice oven spring and almost guarantees that the loaf will not be over-proofed.
- If poor oven spring is a regular occurance, your fermentations may be favoring bacterial growth over yeast growth. Review the starter management tips in Part 2 of this book for help.

Glossary

Like many worthy endeavors, artisan sourdough baking has a robust and entertaining lexicon. **Bolded** terms within the entries have their own stand-alone entry in the glossary.

ACETIC ACID: An acid produced in fermentation by some species of **lactic acid bacteria** that tastes sharp, bright, and tangy. This, for instance, is the acid you taste in vinegar. To increase acetic acid in bread, bacterial population growth can be encouraged by a warm 85–90°F (29–32°C) period in the **bulk fermentation**, followed by a cold **retard** of the loaf during which acetic acid can build up in the dough.

AUTOLYSE: A fancy French term for "soak the flour." Three main activities happen as the flour soaks in the water. First, the **gluten** proteins become wet, so they can start working on their network. Second, any **bran** in the dough from whole grain flour softens and becomes less likely to cut the gluten network during the handling of the dough. The name derives from the third activity: protease and amylase enzymes in the flour are activated to lyse (break down) proteins and starches, respectively, which enhances gluten development.

BACTERIA: In this book, this term refers to **lactic acid bacteria**.

BAGUETTE: A long, narrow loaf baked free form or in a baguette pan.

BAKER'S PERCENTAGE: A standard system of listing bread recipe ingredients that makes it easy to understand quickly the characteristics of the dough. It also greatly simplifies doubling, halving, or making adjustments to the recipe. The weight of flour in the recipe is the standard, and the weight of all other ingredients is expressed as percentages of the weight of the flour. To calculate a baker's percentage, add the weight of all the flour in the recipe to get the total flour in grams. Divide the weight of any other ingredient by the weight of total flour to get the percentage of that ingredient. For example, in a recipe with 500 grams total flour that calls for 10 grams salt, the baker's percentage of salt would be 10 grams / 500 grams = 0.02, which is 2 percent. If that same recipe gave an option to substitute up to 25 percent whole grain flour, the amount of whole grain would be 500 grams total flour × 0.25 = 125 grams whole grain. For sourdough recipes, the amount of water and flour in the starter or levain complicates matters, so we use the **overall baker's percentage**.

BANNETON: A **proofing basket** made of cane or wicker, also called a brotform.

BATARD: An oblong loaf baked free form. Derives from the same Old French word as bastard, being neither a **boule** nor a **baguette**.

BENCH REST: Resting a pre-shaped loaf prior to shaping. This 10- to 30-minute rest allows the gluten to relax enough to prevent tearing or stiffness when the loaf is shaped into its final form.

BOULE: A round loaf baked free form. The French word for bakery, *boulangerie*, comes from this word.

BRAN: The outer part of a kernel of grain that is present in whole grain flours but removed in refined flours such as white flour.

BULK FERMENTATION: Also known as bulk rise or first proof. The first **fermentation** of the dough. In recipes that make multiple loaves, the dough is divided up after the bulk fermentation.

CARAMELIZATION: The browning of sugar that occurs at high temperatures (above 230–360°F; 110–180°C), depending on the type of sugar. During caramelization, sugars are broken down and transformed into new compounds that provide caramel flavor and brown color. This is one of the chemical reactions that causes bread crust to brown while baking; the other is the **Maillard reaction**.

CARBOHYDRATES: Starches and sugars in the flour.

COIL FOLDS: A variation of **stretch and folds** that is performed by lifting the dough in two hands and letting the ends droop toward the bowl like a sleepy cat. The dough is lowered such that one end lays down in the bowl first and the other rests on top of it. When viewed from the side it looks like a spiral. High-hydration doughs respond well to this technique, especially if the bowl has been oiled. This technique is favored by some bakers after the dough has begun to rise because it doesn't deflate the dough as much as stretch and folds.

CRUMB: Everything inside the crust of the bread. An open crumb has large and variable holes and is achieved by higher **hydration levels** and minimizing dough deflating during **shaping**. Lower hydration levels and pressing out gas during shaping lead to a more even, closer crumb with uniform, small holes.

DIASTATIC MALT: A flour made from grains, usually barley or wheat, that have been allowed to sprout. They are dried and milled at a temperature low enough to preserve the enzymes that catalyze the release of maltose from the starches in flour. It is often added to dough to support yeast growth and crust **caramelization**, both of which are enhanced by maltose, and it is included in some commercial white flours. It is optional when called for in recipes and can be purchased or made at home.

DISCARD: Extra sourdough **starter** that is discarded when refreshing the starter. At **refreshment**, some starter must be removed to keep the total amount of starter from increasing. Discarding part of the old starter also helps to rebalance the populations of **yeast** and **bacteria** in the starter to the optimum for bread baking. Discard can be used in sourdough discard recipes. Store it in a covered container in the refrigerator for up to several weeks.

DOUBLE: Referring to dough, **starter**, or **levain**. Rising to twice the volume that it was when first mixed due to gas bubbles produced by the **yeast** being trapped in the **gluten** matrix.

FERMENTATION: The process of transforming the **carbohydrates** in the flour into carbon dioxide gas, alcohol, flavor compounds, or acids by **yeast** and **bacteria** in the **starter**. Fermentation begins happening as soon as starter is added to wet flour and is highly affected by **temperature**. The production of carbon dioxide gas by yeast is what forms the bubbles that make dough rise and prove yeast is alive, so the terms "fermentation," "rise," and "proof" are often used interchangeably in bread making.

GLUTEN: A protein network that forms in dough from the proteins gliadin and glutenin, which are present in wheat flour and many other types of flour. The gluten matrix gives dough its elastic and stretchy quality, holds the carbon dioxide bubbles in the dough through baking, and gives bread its chewy quality. Wheat gluten powder can be added to a recipe to fortify dough, as wheat makes a very strong set of gluten proteins. Some flours make weaker gluten (for example, durum, spelt, rye) or lack gluten (for example, corn, rice, oat).

HOOCH: A liquid produced by a sourdough **starter** when it runs out of food. It contains alcohol, acids, and other flavorful compounds.

HYDRATION LEVEL: The ratio of liquid in the dough with respect to the total weight of the flour in the dough. For example, a recipe with 500 grams of total flour at 80 percent hydration would have 500 grams total flour × 0.8 = 400 grams total liquid in the recipe.

JIGGLE TEST: A test to determine whether the dough is ready during **bulk fermentation** or loaf **proofing**. The dough should be able to jiggle like a bowl of jelly or a water balloon when it's at peak readiness. If it seems bubbly and risen but doesn't jiggle, it is getting very close to being ready. The jiggle test does not apply to cold dough or low-hydration dough, which doesn't jiggle.

LACTIC ACID: A mild acid produced in **fermentation** by **lactic acid bacteria** that tastes smooth, base, and creamy. This is the acid you taste in yogurt, for instance. To increase lactic acid in bread, bacterial population growth can be encouraged by a warm 85–90°F (29–32°C) period in the fermentation.

LACTIC ACID BACTERIA: Bacterial species that make **lactic acid** as their major product when they metabolize (eat) starches and sugars. The wild bacteria that populate a sourdough **starter** belong to this enormous and diverse group.

LAG PHASE: Referring to **yeast** in the **starter** or dough after new flour and water are incorporated. During this phase, yeast enters a metabolic state in which adjustment to the new environment occurs and growth and **fermentation** are slow. Lag phase will be short if the starter recently has been refreshed, and it will be longer the more time has passed since **refreshment**.

LEVAIN: A separate offshoot **refreshment** of the **starter** built in a different container expressly to mix into dough and to bake. There are many advantages to building a levain instead of using regular starter in a recipe, including flexibility and dependability. A levain can be made to any size needed for a recipe, in any proportion desired, using any kind of flour, and can be grown at whichever temperature is desired for the specific goals of the recipe. Since it is freshly grown, a levain will perform optimally in a recipe. The original starter, being kept separate, will be unchanged by these amendments.

MAILLARD REACTION: A chemical reaction that occurs when cooking foods that results in browning and intense flavor development, such as in searing, roasting, or toasting. The temperature must be at least 280°F (140°C) and the process involves transformation of amino acids and sugars into new compounds. This is one of the chemical reactions that causes bread crust to brown while baking; the other is **caramelization**.

OVEN SPRING: The delightful extra rise of the dough after it is placed in the oven. It is caused in large part by a final burst of **yeast** activity. This is what causes the bread to swell where the crust has been scored. There is less oven spring if the loaf was **over-proofed** or if the **bulk fermentation** went way too far, because the yeast ran out of food before the loaf got to the oven. There can be less oven spring if the loaf wasn't scored to let it expand in the oven. There is generally less oven spring the more whole grain flour is used, so you shouldn't expect a lot in a 100 percent whole grain loaf. Baking a retarded loaf straight from the refrigerator can increase oven spring as can **under-proofing** the loaf.

OVERALL BAKER'S PERCENTAGE: This is a version of the **baker's percentage** that is most useful to sourdough baking because this calculation includes the flour and water that were in the **starter**. To calculate, add all the dry flour in the recipe plus half the weight of the starter (assuming the starter is kept at a **hydration level** of 100 percent) to get total flour in grams. Divide all other ingredients by the total flour to get the percentage. For hydration, include half the weight of the starter for the total water. For a recipe that calls for 450 grams flour, 300 grams water, and 100 grams starter, the total flour is 450 grams flour + (½ × 100 grams starter) = 500 grams total flour. The total water is 300 grams water + (½ × 100 grams starter) = 350 grams. The hydration level is 350 grams total water / 500 grams total flour = 0.7 = 70%.

OVER-PROOFED OR OVER-FERMENTED: Used to describe dough that has begun to deteriorate because it has run out of food or is overpopulated by **bacteria** and underpopulated by **yeast**. Many different problems can cause dough to over-ferment, including **resting** too long, using a starter that needs **refreshment**, resting at **temperatures** over 80°F (27°C) for more than a brief period, or using a starter that is incapable of **doubling** in 3 hours. Over-fermented dough is usually loose and slack, smells acidic, and fails to rise or falls upon baking.

PARCHMENT PAPER: A seemingly magical baking paper that cannot catch fire in the oven up to 550°F (288°C) and to which baked goods do not stick unless they were very wet before baking. Using parchment paper reduces dishwashing and it's compostable. It is easy to move your loaves around and to adjust them when they are on parchment liners. There is very little discernible difference between the crust of bread baked directly on a stone or pan and bread that had a parchment liner. Parchment paper is *not* the same as waxed paper, which will burn and stick permanently to your bread.

POKE TEST: A method for testing dough readiness. A wet or floured finger is poked into the dough and the baker waits to see how long it takes the dough to fill the hole back in. If it fills in slowly over several seconds, the dough is ready.

PRE-FERMENT: Dough that is already fermented that is added to the main dough at mixing. It could be a **starter** or a **levain** in sourdough breads.

PRE-FERMENTED FLOUR: The flour in a recipe that came from a **starter** or **levain**. It can be thought of as a proxy for how much **yeast** was added to the recipe because the flour is used by the yeast to multiply during **fermentation**.

PROOF: Also known as a second rise or final rise. Proofing is when a shaped loaf is left to **rest**, allowing it to **rise**. Proofing is done after the **bulk fermentation** and **shaping** the loaves. **Temperature** is a major influence on how long proofing takes. The word derives from the practice of testing the **yeast** before baking the dough to prove that it is active. Some use the term "first proof" in place of "bulk fermentation."

PROOFING BASKET: A container that supports the loaf during proofing and is removed just before baking the loaf. It can be a basket or colander lined with a floured, smooth cloth, or a special basket designed for bread proofing, such as a **banneton**. It must have vertical sides to support the dough, and it is best if it offers some ventilation.

REFRESHMENT: Also known as feeding, when referring to a **starter**. Refreshment is accomplished by removing some starter and **discarding** it or baking with it, and then mixing in new water and flour with the remaining old starter. This process removes "waste" metabolites while providing fresh food, and it also serves to keep the starter weight consistent over time. The balance of yeast and bacterial populations is re-established and yeast comes out of **lag phase**.

REST: Referring to dough, resting means letting the dough sit undisturbed, usually at a particular **temperature**, during the **autolyse**, **fermentation**, **stretch and folds**, or **shaping** stages. Rest is used in a recipe to allow flour to hydrate, **gluten** to relax, the gluten network to form, fermentation of the dough to occur, and the dough to **rise**. A lot is happening during rest!

RETARDING: Slowing down the **fermentation** and **rise** of the dough by chilling it in the refrigerator. **Yeast** activity and growth slows down, **bacterial** growth nearly ceases, and enzymatic processes are altered at low **temperatures**. Retarding can be done during the **bulk fermentation** or **proofing** of the loaf. It is used for purposes of convenience, flavor, and digestibility. You can throw the dough in the refrigerator to accommodate a schedule, or to save one loaf of a two-loaf recipe to be baked the next day. You can also use chilling to achieve a stronger sour flavor or to increase flavor complexity and digestibility in the bread. At low temperatures, a buildup of particular sugars needed to make **acetic acid** (which are released by yeast) occurs. The bacteria will continue making acids, particularly acetic acid, during chilling, so dough that is retarded will become more sour the longer it is chilled. Usually 3 to 24 hours is used.

RISE: The increase in volume of dough, **starter**, or **levain** as it is being **rested**, **fermented**, **proofed**, or baked. If left undisturbed, eventually the rise will hit a peak height, then the yeast will begin to die off and the dough or starter will deflate.

SCORE: A cut through the **skin** of the loaf just before baking to allow and control **oven spring**.

SEAM SIDE: The part of the dough that will become the bottom crust of the finished loaf. Dough has a polarity once it is pre-shaped. The seam side is disorganized and sticky while the top side (because that's where it will be in the finished loaf) is smooth and less sticky.

SHAPING: Forming dough into loaves. Usually shaping is a three-step process that involves pre-shaping and **resting**, shaping, and then tightening the **skin**.

SKIN: The outer surface of the loaf that must be taut to support the loaf's **rise** during **proofing** and baking.

STARTER: Also known as leaven, mother, culture, or barm. It is a diverse living community of wild yeast and bacteria suspended in a flour-and-water paste that is added to dough to make sourdough bread. Starters are commonly kept at 100 percent hydration, but "stiff starters" are kept at lower hydration levels. For the purposes of this cookbook, all starters are kept at 100 percent hydration.

STARTER MANAGEMENT: The art of maintaining your **starter** in a way that works for your schedule and results in the kind of bread you like. Keeping a consistent pattern to starter **refreshment** and **temperature** is helpful for consistent performance. Cooler storage and prompt refreshment will result in a lower ratio of **bacteria** to **yeast**, over multiple refreshments. Increasing the **starter proportion** in refreshment can increase the ratio of bacteria to yeast at peak rise, as can allowing the starter to starve and produce **hooch** between refreshments or **resting** the starter at 85–90°F (29–32°C).

STARTER PROPORTIONS: Usually expressed as a ratio. It tells you how much to add (in weight) during a starter **refreshment**. For example, 1:1:1 means you would add 1 part **starter** to 1 part water and 1 part flour, by weight. If you want to make 60 grams of starter, you would mix 20 grams of the old starter with 20 grams of water and 20 grams of fresh flour. A ratio of 1:2:2 means to 1 part starter add 2 parts water and 2 parts flour. To decrease the starter proportion in refreshments, add more flour and water than starter, such as 1:2:2 or 1:7:7.

STRETCH AND FOLDS: A technique for developing **gluten** in a high-hydration dough that involves stretching one side of the dough and folding it over the rest. It takes only 15 to 30 seconds to complete a set. **Coil folds** are a variation of this technique.

TARE: The weight of the container that must be subtracted to determine the weight of the contents when measuring on a scale. Most scales have a tare function that "zeros out" the weight of the container, so no subtracting is needed.

TEMPERATURE: Temperature is a major influence on how long **fermentation** and **proofing** take and how much sourdough flavor develops. At higher temperatures (85–90°F; 29–32°C), **yeast** grows a bit more slowly while **bacteria** grow their fastest and produce **lactic acid**. At yeast's ideal growing temperature of 74–78°F (23–26°C), rising happens the fastest and bacteria multiply more slowly. As the temperature goes down, the amount of time needed for rising increases dramatically and the growth of bacteria becomes very slow. At refrigeration temperatures, dough rises very slowly and flavor compounds produced by yeast and bacteria accumulate. In addition, the balance of sugars available in the dough encourage bacteria to produce **acetic acid**. Temperature also has a major influence on the ratio of bacteria to yeast in **starter management**.

TOP SIDE: The part of the dough that will become the top crust of the finished loaf. Dough has a polarity once it is pre-shaped. The seam side is disorganized and sticky while the other side, called the top side because that's where it will be in the finished loaf, is smoother with a less sticky feel.

UNDER-PROOFED OR UNDER-FERMENTED: Used to describe dough that has not been given enough time at the **temperature** needed to build up a sufficient population of **yeast** to raise the dough. It is easy to fix by giving the dough more time or by moving it to the ideal temperature for yeast, 74–78°F (23–26°C). Slight under-proofing can be used to increase **oven spring**, to guard against **over-proofing** during a long **retard** or when using whole grains, or to decrease flavor development in a sourdough.

WINDOWPANE: A method for testing gluten development in dough. A piece of dough is stretched as thin as possible without tearing and then is held up to a light source. If light can be seen through the dough, the dough has a well-developed gluten network.

YEAST: The microscopic fungus responsible for producing the carbon dioxide gas that causes dough to **rise** in bread making. In a sourdough **starter**, yeast works together with **lactic acid bacteria**, with which it has a symbiotic relationship. Wild yeast is often slow at rising dough, leading to a lot of flavor development during the extended **fermentation** time with the many wild **bacteria** species present in the starter. Commercial yeasts, which have been bred and purified to raise bread consistently and quickly without lactic acid bacteria, yield breads with a more delicate or bland flavor.

Sourdough Baking Worksheet

DATE:

RECIPE:

CHANGES AND NOTES:

STARTER CONDITION (LAST REFRESHED? RISEN/ FLOATED? ACIDIC?):

LEVAIN

TIME MIXED, TEMPERATURE:

CONDITION AT USE (RISEN? FLOATS IN WATER? ACIDIC?):

BULK FERMENTATION

TIME DOUGH MIXED:

DOUGH TEMPERATURE AND FEEL (STIFF? STICKY? BATTER-LIKE? DRY?):

AUTOLYSE LENGTH:

DOUGH CHECKS

TIME SINCE MIXING	STRETCH AND FOLDS	DOUGH TEMP. & CONDITION

TIME TURNED OUT:

DOUGH CONDITION AT END

JIGGLINESS:

BUBBLINESS, VOLUME GAIN:

SMELL:

ELASTICITY/STRENGTH:

EXTENSIBILITY/WINDOWPANE:

TOTAL TIME IN BULK FERMENTATION (INDICATE HOW MUCH TIME RETARDED):

SHAPING

PRE-SHAPE, LENGTH OF BENCH REST:

SHAPE:

HOW DID IT GO?:

LOAF PROOFING

PROOFING BASKET, LINER, DUSTING FLOUR:

TIME STARTED, TEMPERATURE:

TIME ENDED, HOW MUCH RISEN:

TOTAL PROOFING TIME (INDICATE HOW MUCH TIME RETARDED):

BAKING

TEMPERATURE AND TIME PREHEAT BEGAN:

STEAM/BAKE SETUP:

SCORING PATTERN:

TIME STARTED BAKING:

TIME COVER/STEAM REMOVED:

TIME FINISHED:

LOAF INTERNAL TEMPERATURE:

TOTAL TIME BAKED (INDICATE HOW MUCH TIME STEAMED):

RESULT

OVEN SPRING:

CRUST:

CRUMB:

TASTE:

NOTES FOR NEXT TIME:

THINGS THAT WENT WELL:

THINGS TO CHANGE:

Bibliography

There are many worthy books on bread, baking, and sourdough, and I have read my fair share. Here I list a few that have stood out and have been useful in my sourdough bread baking adventures.

Alterman, Tabitha. *Wholegrain Baking Made Easy.* A nice explanation of how to treat whole grains to get delicious baked goods out of them. Warning: This book will make you covet a grain mill.

Beranbaum, Rose Levy. *The Bread Bible.* A comprehensive book of bread recipes, with information and techniques delivered in a friendly, helpful, and entertaining voice. An invaluable reference for bread bakers.

Buehler, Emily. *Bread Science: The Chemistry and Craft of Making Bread.* A great book for geeking out on the science behind bread making. The very relatable explanations and clear, friendly diagrams make it super approachable for the layperson.

Calvel, Raymond. *The Taste of Bread.* The French baking expert who wanted to bring back traditional quality French artisan bread after the industrialization of bread making changed the character of breads in France explains the techniques and rationale behind the traditional practices. Currently most known for coining the term "autolyse" to refer to soaking the flour before salting or adding yeast.

Ettinger, John. *Bob's Red Mill Baking Book.* Find out about all their many grains and how to use them.

Forkish, Ken. *Flour Water Salt Yeast: The Fundamentals of Artisan Bread and Pizza.* Lots of tips and explanations for understanding how to control your dough in terms of fermentation and gluten development. The recipes, which are mainly commercially yeasted, revolve around a basic bread method. This book details the top factors in making great bread and thoroughly explains the inverse relationship of time and temperature in dough development and the quest for the "sweet spot." Beautiful photographs.

Lahey, Jim. *My Bread: The Revolutionary No-Work, No-Knead Method.* The person most responsible for the modern no-knead bread revolution explains how to make a variety of fantastic breads with little effort (right up my alley!).

Leader, Daniel, and Lauren Chattman. *Living Bread: Tradition and Innovation in Artisan Bread Making.* A nice survey of bread making throughout human history and an in-depth exploration of modern and historic flours and grains used in various countries. Bakers from a number of countries are profiled, and recipes for their breads are presented.

Reinhart, Peter. *The Bread Baker's Apprentice.* Featuring a wide variety of classic bread recipes and helpful information that's clearly presented, this book also includes enjoyable stories, including the author's interactions with icons of modern bread history.

Robertson, Chad. *Tartine Bread.* A book full of beautiful photographs that reveals in a lengthy recipe how the famously delicious Tartine breads are made. Devotion to wild yeast, beautifully grown grains, and a simple set of ingredients during the author's quest for an ideal loaf is detailed as he tells the story of his evolution as a baker. A close reading gives the reader a good sense of the author's method of handling natural leaven.

"The Science of Sourdough." *robdunnlab.com/projects/science-of-sourdough.* North Carolina State University Department of Applied Ecology. You can check out several fascinating science projects at this website to learn more about what is in a sourdough starter. In their Global Sourdough Project, these researchers have determined the species living in starters collected from people around the world and compiled the information in a fun interactive map.

Wood, Ed, and Jean Wood. *Classic Sourdoughs, Revised: A Home Baker's Handbook.* This is a fun book to read. The couple traveled the world collecting sourdough starters, recipes, and friends, and they offer starters for sale to the public. The book features an interesting discussion of managing sourness in bread.

Acknowledgments

No woman is an island, entirely of herself, to paraphrase John Donne, and I am deeply grateful for my connection to the many people who have contributed to the cookbook in your hands.

This book would never have come about without two brilliant gems—my daughters Orelia and Jenya—asking, “What’s there to eat? Is that bread I smell baking? Can I cut into it yet?,” and cheerfully dispatching as many loaves as my many, many experiments produced. Helping them in this task was the star in my solar system, Jack Newman, always quick to remind us that if it’s worth doing, it’s worth overdoing (in case anyone was waffling on whether to indulge in another slice of bread). I feel incredibly lucky for his rock-solid support and encouragement, always, for any kind of adventure I dream of embarking on. They all helped in so many different ways: taste-testing breads, test-baking and posing as hand model, helping me set up the “barbie photo studio”, taking baking instructions over the phone to finish a recipe for me, waiting patiently in the car with the boogie boards while I photographed the focaccia one last time before heading to the beach, and in general filling in on all the things I neglected while researching, writing and baking.

I want to thank my friends and extended family who provided enthusiastic encouragement. My mom, Louise Waldbillig, always supportive, happily handed over all her bread cookbooks. I’m so grateful to those who test-baked the recipes and encouraged me to publish, especially Jahna Balk and Kenichi Sugihara (above and beyond!) and Michael Waldbillig, an experienced sourdough baker. I’m grateful to Nancy Pearlman, cheerleader and bread sampler extraordinaire, and Susan Chester, who inspires me to seek adventure.

It’s one thing to write a book, but a very different thing to get it out into the world. I want to thank my Literary Agent, Neil Gudovitz, for envisioning a place for this book out there and all his efforts in making it happen. I’m so appreciative that he accomplished this in a way that was painless for me.

I am extremely grateful to the entire team at Countryman. In particular, I am indebted to Ann Triestman and Nicolas Teodoro for taking on this project and for being such fun people to work with throughout the process. Their solid feedback truly elevated the book’s content to its full potential. I thank Diane Durrett for expert copyediting—a Herculean task—and Allison Chi and Nick Caruso for the delightful, beautiful design, as well as Devon Zahn, Jess Murphy, and Isabel McCarthy.

Finally, I am indebted to all the women who lovingly crafted delicious food for me throughout my childhood. The images I hold in my mind of their hands in a bowl or stirring a pot, and the serene joy in their faces when placing beautiful food on the table inspire me whenever I cook.

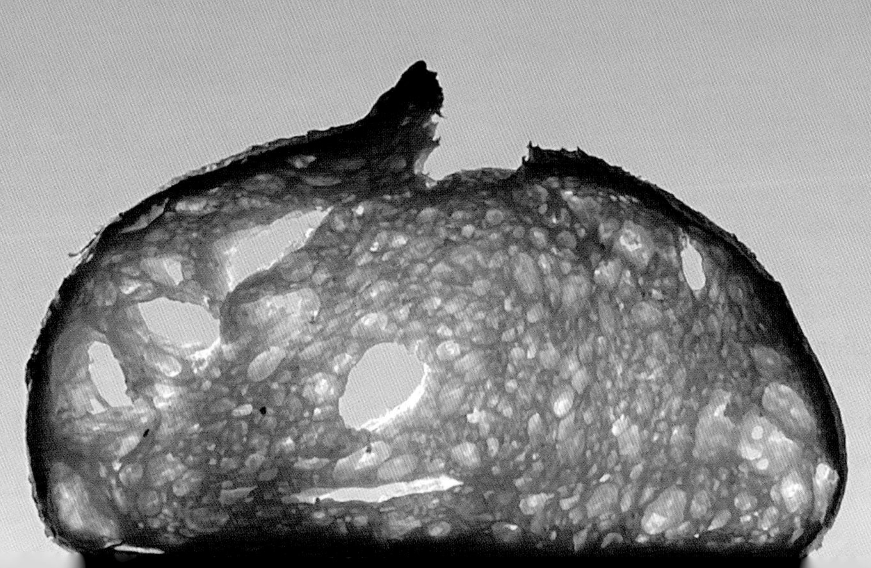

Index

A

B

C

F

G

H

I

J

S

V

W

Y

About the Author

Karyn lives in Northern California with her family, flock of chickens, and devoted vizsla. She enjoys inventing and optimizing recipes for preparing the freshest and most delicious food with the least amount of trouble and waste. (Because there are a lot of other amazing things she likes to do, too!) She began her professional life in editing, then embarked on a career in the sciences after earning her PhD in cellular and molecular biology from the University of Wisconsin, Madison. Research positions at the University of California, Berkeley, brought her to the San Francisco Bay Area, land of 365-day-a-year gardening and unparalleled easy-access natural beauty and creative inspiration. She seriously treasures preparing scrumptious and soul-satisfying food for her family and friends, as well as the simple pleasure of enjoying it together. When not baking she can be found in the garden or on a trail with her boots, bike, or skis and her watercolors.